INDEX

Learning math is like learning a new language. It is probably not among the easiest languages to learn, but it is a language. Math contains a large variety of rules that must be followed and be applied in the correct situation in order to be understood   like a language. Math starts with easy procedures like addition and subtraction, then we use them to build multiplications and divisions. Combining all together we get to exponents. Later we take exponents and move to logarithms, and every step forward is based on the knowledge of its previous step... one more time, like a language. So if it looks like a language, talks like a language, then it is a language!

Same as languages, the most important aspect of its learning is the practice. In order to master the language of math, several hours in problem solving must be spent. The biggest issue with mathematics is; no matter how good we are, we can always encounter a new scenario requiring the slightest change in the solving process, forcing us to waste valuable time trying to get the new answer. For that reason, *Algebra: User's Manual* was created.

Algebra: User's Manual is based in the idea of showing step by step, detail by detail, how to solve many different problems covering as many scenarios as possible. Each of the 150 problems solved was carefully thought, taking into consideration as many scenarios as possible. Simple features are capable of turning a routine problem into a homework nightmare, this book patiently teaches you how to solve each one of them. The book has been designed as a guideline, illustrating each step required to obtain the final answer, explaining every single fact used in the process.

As we reach harder and longer processes, it is common for math books to skip steps already explained in previous chapters. Every now and then, we forget about those past procedures, therefore, we get stuck in harder problems. *Algebra: User's Manual* treats each problem as if it was the first one, making sure not even one single detail is forgotten. In the end, practice is the key for learning math. With *Algebra User's Manual* you will realize how we take the easiest methods learned at first and constantly use them through the entire manual. Watching the same procedure many times helps our brains to get familiarized with them, learning when and where to use them. Eventually you will reach a point in which most of these math rules will pop up automatically in your brain as soon as you need to apply them.

Algebra User's Manual is the first book of the *Math User's Manual* series. Here we will teach you the most important task in math: solving for x. X represents a variable, a missing value that needs to be found in order to solve a problem. Learning how to find variables is a key skill in a wide variety of majors and careers. X could be the strength of a beam, the breakeven point of a product, the time it takes for bacteria to grow and much more. In future releases we will teach you what is x used for up to differential and multivariable calculus.

User manual for the
User's Manual

Algebra User's Manual has been designed to provide aid at every single step of the problem solving process. The book contains several items that will assist our understanding of the process. This section of the book is focused in the explanation of each of these mentioned items. By knowing what each of the elements indicates, we will be able to fully comprehend how to solve for x using algebra.

Number coloring: In most of the steps of the book we will find numbers colored in red and blue. Any number written in RED, represents the action being performed at the current step of the problem. Any number written in BLUE represents the action performed in the previous step. If you have the black and white version there is nothing to worry about, we have designed the colors in such way the tones are easy to differentiate even in black and white.

Moves as division

White bubbles: They include the description of the action realized in the current step.

Arrows: Show the direction of the movement of elements around equations

Friendly assistant: We will use these signs to guide you through the steps of the process. They'll explain what is happening when confusing situations arrive.

Memory card: Sometimes we forget something we learned in the past because we are focused in the current problem. No worries, you can use these ones as reminders.

Foreign assistance: Every concept in math is connected to each other by one way or another. We can't just talk about algebra without using other areas of math, like arithmetic. Each time an outside topic is used, you will see it in these boxes.

Come again?: Every now and then we will face confusing steps and ideas. They might require an extra explanation to make sure we understood them. Well, we have that extra explanation in these items.

Addition, Subtraction, Multiplication and division

Solving for x is the foundation of algebra and it will be an important skill to have in our toolbox during our academic and professional career. Remember x represents a value that needs to be found in order to find the answer of a numeric problem and numeric problems are found in most of science jobs.

In order to solve for x, it must be isolated. By isolation we mean leaving x by itself on one side of the equal sign. Numbers and letter can be moved from one side to the other of the equation by switching it to its opposite element, this is what we call "the opposite rule". Consider that a negative number is the opposite of a positive number, subtraction is the opposite of addition and division is the opposite of multiplication, take a look at the following example:

$$x - 3 = 5$$

In order to fully isolate that x in red, we have to get rid of the negative 3 next to it. The opposite of a negative number is a positive number, so let's add a positive 3:

$$x - 3 + 3 = 5$$

But the main rule of equations says: if we put something on one side of the equal sign, we must do the same in the other side, so a +3 must be included in the right side as well

$$x - 3 + 3 = 5 + 3$$

On the left side the +3 cancels with the -3

$$x - \cancel{3} + \cancel{3} = 5 + 3$$

So our answer is:

$$x = 8$$

But there is a shortcut, instead of doing the addition of the +3 on both sides, we could just do:

$$x - 3 = 5$$

Moves as positive

$$x = 5 + 3$$

$$x = 8$$

Number in red is the operation performed in the current step
Number in blue is the operation performed in the previous step

3

Is the same situation with multiplication and division, let's take a look at the following equation:

$$6x = 18$$

In order to isolate the red x, we need to get rid of the 6 multiplying it. The opposite of multiplication is division so, lets divide by 6:

$$\frac{6x}{6} = 18$$

But don't forget the main rule, what is done on one side of the equal sign, must be done in the other, so we need to divide the right side by 6:

$$\frac{6x}{6} = \frac{18}{6}$$

Both 6 on the left side cancel each other

$$\frac{6\cancel{x}}{\cancel{6}} = \frac{18}{6}$$

Therefore, our final answer is

$$x = 3$$

One more time, the process can be simplified with the shortcut:

$$6x = 18$$

Moves as division

$$x = \frac{18}{6}$$

$$x = 3$$

The shortcut is what we call "the opposite rule", moving elements from one side to the other by switching them to their opposite.

Algebra: User's Manual will use the shortcut through the entire book but every once in a while it will show the long way as a reminder it is the same idea. Now let's proceed to solving problems.

Number in red is the operation performed in the current step
Number in blue is the operation performed in the previous step

Problem 1) $3x - 4 = 14$

$$3x - 4 = 14$$

And then we take care of the number closest from the x

$$3x - 4 = 14$$

Moves as positive

Start by moving the number positioned furthest away from the x. We call it the "furthest away" rule, kinda obvious

$$3x = 14 + 4$$

$$3x = 18$$

Moves as division

$$x = \frac{18}{3}$$

$$x = 6 \text{ »» } \boldsymbol{Answer}$$

Problem 2) $\dfrac{6x}{5} + 12 = 30$

$$\frac{6x}{5} + 12 = 30$$

$$\frac{6x}{5} + 12 = 30$$

The 12 is the number furthest away from the x, so start with it

Moves as negative

$$\frac{6x}{5} = 30 - 12$$

Moves as multiplication

Over here it doesn't really matter which one we do first. The 5 and the 6 have the same proximity to x

$$6x = 18 \cdot 5$$

Moves as division

$$x = \frac{90}{6}$$

$$x = 15 \text{ »» } \boldsymbol{Answer}$$

Number in red is the operation performed in the current step
Number in blue is the operation performed in the previous step

Problem 3) $\dfrac{2x-5}{15} = 3$

$$\dfrac{2x-5}{15} = 3$$

$$\dfrac{2x-5}{15} = 3$$

Moves as multiplication

The 15 is dividing everything on the left side. When this happens, the number in the denominator (15 in this case) acts as a glue keeping everything together. We must remove that glue first, so the very first step is to take that 15 and move it to the other side

And now we can do the "furthest away first" thing

$$2x - 5 = 3 \cdot 15$$

Moves as positive

$$2x = 45 + 5$$

Moves as division

$$x = \dfrac{50}{2}$$

$$x = 25 \text{ »» } \textbf{\textit{Answer}}$$

Solving for x: practice problems

1) $7x + 5 = 26$ **2)** $3 - 2x = 23$

3) $5x + 13 = 48$ **4)** $12 + x = -6$

5) $\dfrac{4x}{5} = 8$ **6)** $\dfrac{12x}{7} - 8 = 10$

7) $\dfrac{2x}{9} + 4 = 14$ **8)** $\dfrac{x-1}{2} = 5$

9) $\dfrac{3x+8}{6} = 12$ **10)** $\dfrac{4-5x}{10} = -2$

Number in red is the operation performed in the current step
Number in blue is the operation performed in the previous step

Exponents

The rule of the opposites also applies for exponents; we just have to remember the opposite of an exponent is a root. For example:

$$x^5 = 4$$

Moves as a root

$$x = \sqrt[5]{4}$$

Exponents are the next step after addition/subtraction/multiplication/division in math world. The main idea remains; we must isolate the x in order to solve for it. The process of "furthest away first" it is still applied. Most of the times the exponent is the last element to be moved to the other side but let's take a look at some examples to get practice.

Problem 4) $4x^2 - 10 = 90$

$$4x^2 - 10 = 90$$

$$4x^2 - 10 = 90$$

We must follow the "furthest away" rule, so first we move the -10, then the 4 multiplying and the exponent would be the last one

Moves as positive

$$4x^2 = 90 + 10$$

Moves as division

Remember, whenever a root does not have any number in this spot, we can assume: $\sqrt{x} = \sqrt[2]{x}$

$$x^2 = 100/_4$$

Moves as a root

Square roots always have one positive and one negative answer

$$x = \sqrt{25}$$

$$x = \sqrt[2]{25}$$

$$x = 5 \text{ and } x = -5 \text{ »» } \textbf{\textit{Answer}}$$

Number in red is the operation performed in the current step
Number in blue is the operation performed in the previous step

Problem 5) $2(x-1)^2 = 32$

$$2(x-1)^2 = 32$$

$$2(x-1)^2 = 32$$

Moves as division

Whenever we deal with x inside of a parenthesis we first have to clear the parenthesis. The parenthesis takes the role of the glue this time. We must move the 2 multiplying the parenthesis, then the exponent so the parenthesis is fully isolated

$$(x-1)^2 = {}^{32}/_2$$

Moves as a root

Square roots always have one positive and one negative answer

$$x - 1 = \sqrt{16}$$

$$x - 1 = 4 \qquad\qquad x - 1 = -4$$

Moves as positive   Moves as positive

$$x = 4 + 1 \qquad\qquad x = -4 + 1$$

$$x = 5 \qquad and \qquad x = -3 \text{ »» } Answer$$

Problem 6) $x^3 + 2 = 29$

$$x^3 + 2 = 29$$

$$x^3 + 2 = 29$$

Moves as negative

Take a good look here. There are no parenthesis. The 3 is way closer to the x than the +2, therefore, we need to start with the 2

For odd powered roots (like this one; root three) the answer will always have the same sign as the number inside(positive 27 for this case)

$$x^3 = 29 - 2$$

Moves as a root

$$x = \sqrt[3]{27}$$

$$x = 3 \text{ »» } Answer$$

Number in red is the operation performed in the current step
Number in blue is the operation performed in the previous step

8

Problem 7) $\dfrac{3x^4}{4} - 10 = 20$

$$\dfrac{3x^4}{4} - 10 = 20$$

$$\dfrac{3x^4}{4} - 10 = 20$$

Moves as positive

$$\dfrac{3x^4}{4} = 20 + 10$$

Moves as multiplication

$$3x^4 = 30 \cdot 4$$

Moves as division

$$x^4 = {120}/{3}$$

Moves as a root

Square roots always have one positive and one negative answer

When the root has a long decimal answer, we can leave it as a root

$$x = \sqrt[4]{40}$$

$$x = \sqrt[4]{40} \ and \ x = -\sqrt[4]{40} \ »» \ Answer$$

Exponents: practice problems

1) $x^2 = 16$ **2**) $3x^2 - 5 = 70$ **3**) $\dfrac{x^3}{4} - 6 = 10$

4) $8 + 2x^4 = 116$ **5**) $\dfrac{x^2 - 4}{3} = 8$ **6**) $\dfrac{1 - x^2}{5} = 2$

Number in red is the operation performed in the current step
Number in blue is the operation performed in the previous step

Roots

Roots are one of the most feared topics in algebra and why not? Their concept is quite confusing; we are looking for a number that multiplied itself a certain amount of times is equal to the number inside of the root   wait what?? Yes, that is what we know about roots. But actually a root is just an exponent written in fraction form:

$$\sqrt[2]{x} = x^{\frac{1}{2}} \;\; or \;\; \sqrt[5]{x^4} = x^{\frac{4}{5}}$$

This technique will come handy in algebra, trigonometry, calculus and all other math subjects we may have to face in our academic career:

$$\sqrt[n]{x^m} = x^{\frac{m}{n}}$$

This equation applies for all the roots in math world. Just keep in mind the exponent of the number inside of the root is the numerator of the fraction and the number outside of the root is the denominator of the fraction.

As mentioned in the *exponent* section; roots are the opposite of exponents, meaning:

$$\sqrt[3]{x} = 4$$

Moves as exponent

$$x = 4^3 \rightarrow x = 64$$

Let's view some problems to get some practice.

Problem 8) $\sqrt{x} - 2 = 3$

$$\sqrt{x} - 2 = 3$$

$$\sqrt{x} - 2 = 3$$

Same idea as with exponents, we have to first isolate the root

Moves as positive

Remember, whenever a root does not have any number in this spot, we can assume:
$$\sqrt{x} = \sqrt[2]{x}$$

$$\sqrt[2]{x} = 3 + 2$$

Moves as exponent

$$x = 5^2$$

Here we have a tricky concept that doesn't affect our answer, but we will discuss it in the next problem because we have more space there

$$x = 25 \; »» \; \textbf{\textit{Answer}}$$

Number in red is the operation performed in the current step
Number in blue is the operation performed in the previous step

Problem 9) $3\sqrt{x-2} = 18$

$$3\sqrt{x-2} = 18$$

$$\mathbf{3}\sqrt{x-2} = 18$$

Moves as division

The x is inside of the root but there is a number multiplying it, that number must be moved first

Remember, whenever a root does not have any number in this spot, we can assume:
$\sqrt{x} = \sqrt[2]{x}$

$$\sqrt[2]{x-2} = {}^{18}\!/_3$$

Moves as exponent

Weird math concept alert: Every time we elevate a number by an even power, we introduce a positive and a negative answer (like square roots). It doesn't really matter at this level, but it will be important to remember this in more advanced math

$$x - \mathbf{2} = 6^2$$

Moves as positive

$$x = 36 + 2$$

$$x = \mathbf{38} \text{ »» } \textit{Answer}$$

Problem 10) $\sqrt{3x-2} = 5$

$$\sqrt{3x-2} = 5$$

$$\sqrt[2]{3x-2} = 5$$

Moves as exponent

Everything on the left side is inside of the root, in this case, the root acts as a glue and needs to be removed first

$$3x - \mathbf{2} = 5^2$$

Moves as positive

Just to show a bit more about that weird math concept. Over here, technically we would have to use -5 and +5. The reason is, because -5 squared and +5 squared they are both equal to +25. As we said, it does not make much of a difference for now

$$\mathbf{3}x = 25 + 2$$

Moves as division

$$x = {}^{27}\!/_3$$

$$x = \mathbf{9} \text{ »» } \textit{Answer}$$

Number in red is the operation performed in the current step
Number in blue is the operation performed in the previous step

11

Problem 11) $\frac{\sqrt{x+5}}{3} = 1$

$$\frac{\sqrt{x+5}}{3} = 1$$

$$\frac{\sqrt{x+5}}{3} = 1$$

Moves as multiplication

The 3 is dividing outside of the root, meaning we first have to move it to the other side before we can take care of the root. The 3 is the glue

$$\sqrt[2]{x+5} = 1 \cdot 3$$

Moves as exponent

$$x + 5 = 3^2$$

Moves as negative

$$x = 9 - 5$$

$$x = 4 \text{ »» } Answer$$

Number in red is the operation performed in the current step
Number in blue is the operation performed in the previous step

12

Problem 12) $\sqrt{\dfrac{x+5}{3}} = 1$

$$\sqrt{\dfrac{x+5}{3}} = 1$$

$$\sqrt[2]{\dfrac{x+5}{3}} = 1$$

Moves as exponent

Very similar as the previous problem, but now the 3 is inside, meaning we MUST start with the root. Now the root is the glue

$$\dfrac{x+5}{3} = 1^2$$

Moves as multiplication

$$x + 5 = 1 \cdot 3$$

Moves as negative

$$x = 3 - 5$$

$$x = -2 \text{ »» } \textbf{\textit{Answer}}$$

Problem 13) $\sqrt[5]{x} + 10 = 12$

$$\sqrt[5]{x} + 10 = 12$$

$$\sqrt[5]{x} + 10 = 12$$

Moves as negative

First move everything NOT inside of the root to the other side

That weird math concept does not apply here because we have an odd exponent. That thing only happens with even exponent. Please remember, it does not make much of a difference now, but it Will be useful in the far future.

$$\sqrt[5]{x} = 12 - 10$$

Moves as exponent

$$x = 2^5$$

$$x = 32 \text{ »» } \textbf{\textit{Answer}}$$

Number in red is the operation performed in the current step
Number in blue is the operation performed in the previous step

Problem 14) $\sqrt[6]{x} + 9 = 8$

$$\sqrt[6]{x} + 9 = 8$$

$$\sqrt[6]{x} + 9 = 8$$

Moves as negative

$$\sqrt[6]{x} = 8 - 9$$

Moves as exponent

$$x = (-1)^6$$

$$x = 1 \text{ »» } \textbf{Answer}$$

Problem 15) $\sqrt[3]{x - 2} = 5$

$$\sqrt[3]{x - 2} = 5$$

$$\sqrt[3]{x - 2} = 5$$

Moves as exponent

$$x - 2 = 5^3$$

Moves as positive

$$x = 125 + 2$$

$$x = 127 \text{ »» } \textbf{Answer}$$

Number in red is the operation performed in the current step
Number in blue is the operation performed in the previous step

Problem 16) $\sqrt[4]{x^2 + 5} = 3$

$$\sqrt[4]{x^2 + 5} = 3$$

$$\sqrt[4]{x^2 + 5} = 3$$

Moves as exponent

$$x^2 + 5 = 3^4$$

Moves as negative

$$x^2 = 81 - 5$$

Moves as a root

Remember, whenever a root does not have any number in this spot, we can assume: $\sqrt{x} = \sqrt[2]{x}$

Don't forget square roots always have two answers; a positive and a negative

$$x = \sqrt[2]{76}$$

$$x = \sqrt{76} \quad and \quad x = -\sqrt{76}$$
Answers

Roots: practice problems

1) $\sqrt{x} - 4 = 5$ **2)** $\sqrt[3]{x} + 10 = 2$ **3)** $\sqrt{x + 10} + 1 = 7$

4) $\sqrt[4]{2x - 1} + 4 = 3$ **5)** $\dfrac{\sqrt[3]{x}}{6} = 36$ **6)** $\dfrac{\sqrt{2x - 1}}{3} = 2$

7) $\dfrac{\sqrt{2x} - 1}{3} = 2$ **8)** $\sqrt{\dfrac{2x - 1}{3}} = 2$

Number in red is the operation performed in the current step
Number in blue is the operation performed in the previous step

15

Parenthesis

Parenthesis are math's organizing crew. We use them to group elements within an equation, to highlight their importance, specify their order or simply to make the equation look nice. When solving for equations involving parenthesis we have two alternatives, eliminate whatever is adjacent to them (multiplying numbers, signs and exponents) or distribute them. Both methods will be shown in this section but distributing the parenthesis will be given priority.

Problem 17) $3(5x - 2) = 10$

$$3(5x - 2) = 10$$

$$3(5x - 2) = 10$$

Distribute the 3 inside of the parenthesis

We distributed the parenthesis so we could get rid of it and now we have a normal equation we know how to solve

$$3 \cdot 5x - 3 \cdot 2 = 10$$

$$15x - 6 = 10$$

Don't forget the order of operations:

MULTIPLICATION FIRST!!

Moves as positive

$$15x = 10 + 6$$

Moves as division

$$x = \frac{16}{15}$$

$$x = \frac{16}{15} \text{ »» } \textbf{\textit{Answer}}$$

Alternative method

By moving the 3 to the other side we are automatically eliminating the parenthesis

$$3(5x - 2) = 10$$

Moves as division

$$5x - 2 = \frac{10}{3}$$

Moves as positive

Number in red is the operation performed in the current step
Number in blue is the operation performed in the previous step

$$5x = \frac{10}{3} + 2$$

$$5x = \frac{16}{3}$$

Moves as division

Apply this rule for adding fractions:
$$\frac{A}{B} + \frac{C}{D} = \frac{A \cdot D + B \cdot C}{B \cdot D}$$

So
$$\frac{10}{3} + \frac{2}{1} = \frac{10 \cdot 1 + 3 \cdot 2}{3 \cdot 1}$$

$$x = \frac{16}{3 \cdot 5}$$

$$x = \frac{16}{15} \text{ »» } \textbf{Answer}$$

Problem 18) $7(2x + 4) + 10 = -18$

$$7(2x + 4) + 10 = -18$$

$$7(2x + 4) + 10 = -18$$

Distribute the 7 inside of the parenthesis

$$7 \cdot 2x + 7 \cdot 4 + 10 = -18$$

$$14x + 28 + 10 = -18$$

Add the two terms without x

We have two negatives, we must add them together and keep their sign, therefore:

$$-18 - 38 = -56$$

If only one of the elements of the fraction (in this case the numerator) has a negative, then the entire answer is negative

$$14x + 38 = -18$$

$$14x + 38 = -18$$

Moves as negative

$$14x = -18 - 38$$

Moves as division

$$x = \frac{-56}{14}$$

$$x = -4 \text{ »» } \textbf{Answer}$$

Number in red is the operation performed in the current step
Number in blue is the operation performed in the previous step

Alternate method

$$7(2x + 4) + 10 = -18$$

Moves as negative

$$7(2x + 4) = -18 - 10$$

We have two negatives, we must add them together and keep their sign, therefore:

$$-4 - 4 = -8$$

Moves as division

$$2x + 4 = \frac{-28}{7}$$

Moves as negative

$$2x = -4 - 4$$

Moves as division

$$x = \frac{-8}{2}$$

$$x = -4 \text{ »» } \boldsymbol{Answer}$$

Problem 19) $-2(x - 4) - 6 = -12$

$$-2(x - 4) - 6 = -12$$

$$-2(x - 4) - 6 = -12$$

Distribute the -2 inside of the parenthesis

Two negative numbers are being multiplied. Negative times a negative gives a positive:

$$(-2) \cdot (-4) = +8$$

$$-2 \cdot x - 2 \cdot -4 - 6 = -12$$

$$-2x + 8 - 6 = -12$$

Add the two terms without x

$$-2x + 2 = -12$$

Number in red is the operation performed in the current step
Number in blue is the operation performed in the previous step

$$-2x + 2 = -12$$

Moves as negative

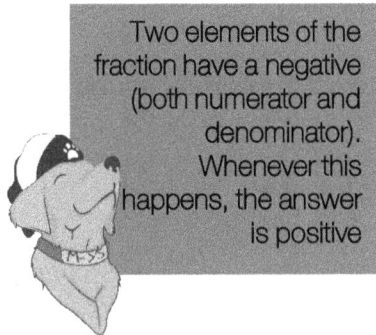

Two elements of the fraction have a negative (both numerator and denominator). Whenever this happens, the answer is positive

$$-2x = -12 - 2$$

Moves as division

$$x = \frac{-14}{-2}$$

$$x = 7 \text{ »» } \textbf{\textit{Answer}}$$

Problem 20) $5(2x^2 + 4) = 30$

$$5(2x^2 + 4) = 30$$

$$5(2x^2 + 4) = 30$$

Distribute the 5 inside of the parenthesis

$$5 \cdot 2x^2 + 5 \cdot 4 = 30$$

$$10x^2 + 20 = 30$$

Moves as negative

$$10x^2 = 30 - 20$$

Moves as division

$$x^2 = \frac{10}{10}$$

Moves as a root

$$x = \sqrt{1}$$

$$x = 1 \text{ and } x = -1 \text{ »» } \textbf{\textit{Answer}}$$

Number in red is the operation performed in the current step
Number in blue is the operation performed in the previous step

Parenthesis: practice problems

1) $4(x + 12) = 36$ **2**) $3(8 - 5x) = 9$ **3**) $-2(11x + 8) = 40$

4) $-(5 - x) = -4$ **5**) $10\left(\dfrac{4x + 8}{6}\right) = 140$

No fractions involved

Here is when the language of math starts getting more complicated. We are including more letters in a subject supposedly related to just numbers. Hey but let's not panic yet. For the following examples we will try to put together all the elements of the equation containing the same variable in one group and all other elements without it in another group. After that, our equation will become a lot easier.

The rule is very straight forward. If the elements have the same variable and the variable has the same exponent, we can add them together, if either the variable or its exponent is different, we can NOT add them. Like this:

$$4x + 6x - 2x \rightarrow 8x$$

In the above example, all the terms have an x raised to the power of 1, meaning we just add the numbers together and keep the x.

But if we look at:

$$6x + 7x^2 - 5x^3 \rightarrow 6x + 7x^2 - 5x^3$$

Even though they have the same variable x, their exponent is different so we can't join them together. This section contains plenty of examples. Let's take a look at them in order to make this rule easier to understand.

Problem 21) $5x + 2x = 14$

Both terms have the same x and the x have the same exponent, which is why we can add them.

$5x + 2x = 7x$

$$5x + 2x = 14$$

$$5x + 2x = 14$$

Combine like terms

And now we have a simple equation like the ones we are used to

$7x = 14$

$$7x = 14$$

Moves as division

$$x = \frac{14}{7}$$

$$x = 2 \text{ »» } Answer$$

Number in red is the operation performed in the current step
Number in blue is the operation performed in the previous step

21

Problem 22) $8x - 3x = 25$

$$8x - 3x = 25$$

$$8x - 3x = 25$$

Combine like terms

$$5x = 25$$

$$5x = 25$$

Moves as division

$$x = \frac{25}{5}$$

$$x = 5 \text{ »» } \textit{Answer}$$

Problem 23) $2x - 4x = 16$

$$2x - 4x = 16$$

$$2x - 4x = 16$$

And now we have a simple equation like the ones we are used to

Combine like terms

$$-2x = 16$$

$$-2x = 16$$

Moves as division

In math when we say "combine" we mean "apply addition or subtraction based on the symbols"
We have a positive and a negative, we must subtract them and keep the sign of the larger number, in this case, the negative 4:
$$2 - 4 = -2$$

$$x = \frac{16}{-2}$$

$$x = -8 \text{ »» } \textit{Answer}$$

Number in red is the operation performed in the current step
Number in blue is the operation performed in the previous step

Problem 24) $7x + 5 + 3x + 8 = 33$

$$7x + 5 + 3x + 8 = 33$$

$$7x + 5 + 3x + 8 = 33$$

We are trying to simplify this equation, in order to do that, let's start by adding all the elements containing an x:
$7x + 3x = 10x$

| Combine elements with x |

$$10x + 5 + 8 = 33$$

And now add the elements not containing any x:
$5 + 8 = 13$

| Combine elements without x |

$$10x + 13 = 33$$

$$10x + 13 = 33$$

| Moves as negative |

There we go, an equation in a form we already know how to solve

$$10x = 33 - 13$$

| Moves as division |

$$x = \frac{20}{10}$$

$$x = 2 \text{ »» } \textbf{Answer}$$

Problem 25) $2x - 4x = 16$

$$7x + 10 - 4x - 8 = 14$$

$$7x + 10 - 4x - 8 = 14$$

Again, we worry first about the elements with x, following the rules of algebra:

$7x - 4x = 3x$

| Combine elements with x |

$$3x + 10 - 8 = 14$$

Then, take care of anything not having an x:

$10 - 8 = 2$

| Combine elements without x |

$$3x + 2 = 14$$

Number in red is the operation performed in the current step
Number in blue is the operation performed in the previous step

$$3x + 2 = 14$$

Moves as negative

$$3x = 14 - 2$$

Moves as division

$$x = \frac{12}{3}$$

$$x = 4 \text{ »» } \textbf{Answer}$$

Problem 26) $-12x + 25 + 9x - 30 = -20$

$$-12x + 25 + 9x - 30 = -20$$

$$-12x + 25 + 9x - 30 = -20$$

Same pattern, group all the elements with x first:
$9x - 12x = -3x$

Combine elements with x

$$-3x + 25 - 30 = -20$$

Afterwards, we group elements without x:

$25 - 30 = -5$

Combine elements without x

$$-3x - 5 = -20$$

$$-3x - 5 = -20$$

Moves as positive

$$-3x = -20 + 5$$

Moves as division

$$x = \frac{-15}{-3}$$

$$x = 5 \text{ »» } \textbf{Answer}$$

Number in red is the operation performed in the current step
Number in blue is the operation performed in the previous step

Problem 27) $3x + 6 = 2x + 10$

$$3x + 6 = 2x + 10$$

$$3x + 6 = 2x + 10$$

Moves as negative

We have x's on both sides of the equation. Variables like to be with their own kind so, our very first step is to group them all at the same side of The equal sign.

$$3x - 2x + 6 = 10$$

$$3x - 2x + 6 = 10$$

Now on the same side, we can combine them: $3x - 2x = 1x = x$

Combine elements with x

$$x + 6 = 10$$

$$x + 6 = 10$$

Moves as negative

$$x = 10 - 6$$

$$x = 4 \text{ »» } \textbf{\textit{Answer}}$$

Problem 28) $7x + 12 = 9x - 34$

$$7x + 12 = 9x - 34$$

$$7x + 12 = 9x - 34$$

Moves as negative

Always start by grouping the x's in the same side of equation

$$7x - 9x + 12 = -34$$

$$7x - 9x + 12 = -34$$

Now on the same side, we can combine them: $7x - 9x = -2x$

Combine elements with x

$$-2x + 12 = -34$$

Number in red is the operation performed in the current step
Number in blue is the operation performed in the previous step

$$-2x + 12 = -34$$

Moves as negative

$$-2x = -34 - 12$$

Moves as division

$$x = \frac{-46}{-2}$$

$$x = 23 \text{ »» } \textbf{\textit{Answer}}$$

Problem 29) $2x + 6 = 18 - 4x$

$$2x + 6 = 18 - 4x$$

$$2x + 6 = 18 \;\; -4x$$

Moves as positive

Unite the x's together! For the good of the nation, I mean equation!

$$2x + 4x + 6 = 18$$

$$2x + 4x + 6 = 18$$

Now on the same side, we can combine them:
$2x + 4x = 6x$

Combine elements with x

$$6x + 6 = 18$$

$$6x + 6 = 18$$

Moves as negative

$$6x = 18 - 6$$

Moves as division

$$x = \frac{12}{6}$$

$$x = 2 \text{ »» } \textbf{\textit{Answer}}$$

Number in red is the operation performed in the current step
Number in blue is the operation performed in the previous step

Problem 30) $-8x + 5 = -x - 30$

$$-8x + 5 = -x - 30$$

$$-8x + 5 = -x - 30$$

Moves as positive

Last reminder about putting all the x's in the same side

$$-8x + x + 5 = -30$$

$$-8x + x + 5 = -30$$

Combine elements with x

And last one about combining them:

$$x - 8x = -7x$$

$$-7x + 5 = -30$$

$$-7x + 5 = -30$$

Moves as negative

$$-7x = -30 - 5$$

Moves as division

$$x = \frac{-35}{-7}$$

$$x = 5 \;\text{»» } \textbf{\textit{Answer}}$$

Number in red is the operation performed in the current step

Number in blue is the operation performed in the previous step

Problem 31) $5(x + 4) = 8(2x + 8)$

$$5(x + 4) = 8(2x + 8)$$

Whenever a number is multiplying a parenthesis, it is easier if we start by distributing that number

$$5(x + 4) = 8(2x + 8)$$

Distribute the 5 and the 8 within their respective parenthesis

$$5 \cdot x + 5 \cdot 4 = 8 \cdot 2x + 8 \cdot 8$$

$$5x + 20 = 16x + 64$$

$$5x + 20 = 16x + 64$$

Ok, one more reminder about putting all the x's in the same side

Moves as negative

$$5x - 16x + 20 = 64$$

$$-11x + 20 = 64$$

Moves as negative

$$-11x = 64 - 20$$

Only one component of the fraction is negative (the denominator in this case) so the final answer is negative

Moves as division

$$x = \frac{44}{-11}$$

$$x = -4 \text{ »» Answer}$$

Number in red is the operation performed in the current step
Number in blue is the operation performed in the previous step

28

Problem 32) $3(10 - 4x) = 2(2x + 55)$

$$3(10 - 4x) = 2(2x + 55)$$

$$3(10 - 4x) = 2(2x + 55)$$

Distribute the 3 and the 2 within their respective parenthesis

$$3 \cdot 10 - 3 \cdot 4x = 2 \cdot 2x + 2 \cdot 55$$

$$30 - 12x = 4x + 110$$

$$30 - 12x = 4x + 110$$

As usual, let's distribute the parenthesis first

Moves as negative

Sorry, there are no more reminders about putting all the x's in the same side

$$30 - 12x - 4x = 110$$

$$30 - 16x = 110$$

Moves as negative

$$-16x = 110 - 30$$

Moves as division

$$x = \frac{80}{-16}$$

$$x = -5 \text{ »» } \boldsymbol{Answer}$$

Number in red is the operation performed in the current step
Number in blue is the operation performed in the previous step

29

Problem 33) $10(2x - 4) = 8(5 - 4x)$

$$10(2x - 4) = 8(5 - 4x)$$

$$10(2x - 4) = 8(5 - 4x)$$

Distribute the 10 and the 8 within their respective parenthesis

$$10 \cdot 2x - 10 \cdot 4 = 8 \cdot 5 - 8 \cdot 4x$$

$$20x - 40 = 40 - 32x$$

$$20x - 40 = 40 - 32x$$

Moves as positive

$$20x + 32x - 40 = 40$$

$$20x + 32x - 40 = 40$$

Moves as positive

Always simplify your fractions as much as possible:

$$\frac{80 \div 4}{52 \div 4} = \frac{20}{13}$$

When the fraction gives a decimal answer, we can just leave it as a fraction

$$52x = 40 + 40$$

Moves as division

$$x = \frac{80}{52}$$

$$x = \frac{20}{13} \quad »» \textbf{\textit{Answer}}$$

Number in red is the operation performed in the current step

Number in blue is the operation performed in the previous step

<u>Different x's same power: practice problems</u>

1) $5x + 10 - 2x + 4 = 20$ **2)** $15x^2 - 9 = 21 + 10x^2$

3) $x^4 + 3x^4 + 16 - 10x^4 = 4$ **4)** $2x + 8 - 3x - 7 = 2$

5) $2\sqrt{x} + 11 = \sqrt{x} + 17$ **6)** $3(4 - 10x) = -8(12x + 1)$

7) $6(7 + 2x) = 5(3x + 9)$ **8)** $-2(x + 5) = -(30 - 3x)$

Number in red is the operation performed in the current step
Number in blue is the operation performed in the previous step

31

Fractions

If we follow the analogy about math being a language, this is the moment when we start dealing with all the different past tenses of verbs. Fractions is that part of math no one likes to study but unfortunately we need to if we want to fully understand them. Don't worry, here we will take our time to explain each detail related to fractions operations until the problems are solved. The main idea is the same, group all the x's into the same side of the equal sign, but now we are also going to try to keep them within the same fraction. Two important rules to remember:

- Get rid of the fractions as soon as possible

- We don't like x's in the denominator, get them out of there quickly

As usual, lets view some solved examples to better understand the concept

Problem 34) $\dfrac{3}{x-1} = \dfrac{2}{2x-6}$

$$\frac{3}{x-1} = \frac{2}{2x-6}$$

Rule n° 2 says we don't like x's in the denominator, so let's move them out of there

$$\frac{3}{x-1} \diagdown \frac{2}{2x-6}$$

Move each denominator as a multiplication to the other side

Now distribute your parenthesis. Is the problem looking familiar?

$$3(2x-6) = 2(x-1)$$

$$3(2x-6) = 2(x-1)$$

Distribute the 3 and the 2 within their respective parenthesis

$$3 \cdot 2x - 3 \cdot 6 = 2 \cdot x - 2 \cdot 1$$

$$6x - 18 = 2x - 2$$

$$6x - 18 = 2x - 2$$

Don't forget to add the elements having the same x

Moves as negative

$$6x - 2x - 18 = -2$$

$$4x - 18 = -2$$

Moves as positive

Sorry, we told you, no more reminders about putting all x's in the same side

$$4x = -2 + 18$$

Moves as division

Number in red is the operation performed in the current step
Number in blue is the operation performed in the previous step

$$x = \frac{16}{4}$$

$$x = 4 \text{ »» } Answer$$

Problem 35) $\frac{6}{3-x} = \frac{5}{10x+2}$

$$\frac{6}{3-x} = \frac{5}{10x+2}$$

Again, rule n° 2. Don't leave x's in the Denominator for too long

$$\frac{6}{3-x} \quad \frac{5}{10x+2}$$

Move each denominator as a multiplication to the other side

One more time, distribute the parenthesis

$$6(10x+2) = 5(3-x)$$

$$6(10x+2) = 5(3-x)$$

Distribute the 6 and the 5 within their respective parenthesis

$$6 \cdot 10x + 6 \cdot 2 = 5 \cdot 3 - 5 \cdot x$$

$$60x + 12 = 15 - 5x$$

$$60x + 12 = 15 - 5x$$

Moves as positive

Sorry, we told you, no more reminders about putting all x's in the same side

$$60x + 5x + 12 = 15$$

$$65x + 12 = 15$$

Moves as negative

$$65x = 15 - 12$$

Moves as division

$$x = \frac{3}{65}$$

$$x = \frac{3}{65} \text{ »» } Answer$$

Number in red is the operation performed in the current step
Number in blue is the operation performed in the previous step

Problem 36) $\frac{2}{x^3-1} = \frac{10}{x^3+5}$

$$\frac{2}{x^3 - 1} = \frac{10}{x^3 + 5}$$

$$\frac{2}{x^3 - 1} \diagdown \frac{10}{x^3 + 5}$$

Move each denominator as a multiplication to the other side

$$2(x^3 + 5) = 10(x^3 - 1)$$

$$2(x^3 + 5) = 10(x^3 - 1)$$

Distribute the 2 and the 10 within their respective parenthesis

$$2 \cdot x^3 + 2 \cdot 5 = 10 \cdot x^3 + 10 \cdot -1$$

$$2x^3 + 10 = 10x^3 - 10$$

$$2x^3 + 10 = 10x^3 - 10$$

Moves as negative

$$2x^3 - 10x^3 + 10 = -10$$

$$-8x^3 + 10 = -10$$

Elements having the same x with the same exponent can be combined together:

$$2x^3 - 10x^3 = -8x^3$$

Moves as negative

$$-8x^3 = -10 - 10$$

Moves as division

Double negative in the fraction, this means our answer will be positive

$$x^3 = \frac{-20}{-8}$$

$$x^3 = \frac{5}{2}$$

Moves as a root

Just the simplified version of the fraction:

$$\frac{20 \div 4}{8 \div 4} = \frac{5}{2}$$

When the root has a long decimal answer, we can leave it as a root

$$x = \sqrt[3]{5/2}$$

$$x = \sqrt[3]{5/2} \quad \text{»» Answer}$$

Number in red is the operation performed in the current step
Number in blue is the operation performed in the previous step

Problem 37) $\dfrac{12}{x^2+5} = \dfrac{4}{2x^2-6}$

$$\frac{12}{x^2 + 5} = \frac{4}{2x^2 - 6}$$

$$\frac{12}{x^2 + 5} \bowtie \frac{4}{2x^2 - 6}$$

Move each denominator as a multiplication to the other side

$$12(2x^2 - 6) = 4(x^2 + 5)$$

$$12(2x^2 - 6) = 4(x^2 + 5)$$

Distribute the 12 and the 4 within their respective parenthesis

$$12 \cdot 2x^2 + 12 \cdot -6 = 4 \cdot x^2 + 4 \cdot 5$$

$$24x^2 - 72 = 4x^2 + 20$$

$$24x^2 - 72 = 4x^2 + 20$$

Moves as negative

$$24x^2 - 4x^2 - 72 = 20$$

Elements having the same x with the same exponent can be combined together:

$$24x^2 - 4x^2 = 20x^2$$

$$20x^2 - 72 = 20$$

Moves as positive

$$20x^2 = 20 + 72$$

Moves as division

$$x^2 = \frac{92}{20}$$

Just the simplified version of the fraction:

$$\frac{92 \div 4}{20 \div 4} = \frac{18}{5}$$

$$x^2 = {}^{23}\!/_5$$

Moves as a root

When the root has a long decimal answer, we can leave it as a root

$$x = \sqrt[2]{{}^{23}\!/_5}$$

$$x = \sqrt{{}^{23}\!/_5} \; and \; x = -\sqrt{{}^{23}\!/_5} \;»» \; Answer$$

Number in red is the operation performed in the current step
Number in blue is the operation performed in the previous step

Problem 38) $\frac{x}{3} + x = 5$

$$\frac{x}{3} + x = 5$$

Now we have a nice and easy problem involving just one single x

$$\frac{x}{3} + x = 5$$

Combine the two elements with x

$$\frac{4x}{3} = 5$$

The rule for addition of fractions applies here:
$$\frac{A}{B} + \frac{C}{D} = \frac{A \cdot D + B \cdot C}{B \cdot D}$$

$$\frac{x}{3} + \frac{x}{1} = \frac{x \cdot 1 + 3 \cdot x}{3 \cdot 1} =$$
$$\frac{x + 3x}{3} = \frac{4x}{3}$$

$$\frac{4x}{3} = 5$$

Moves as multiplication

$$4x = 5 \cdot 3$$

Moves as division

$$x = \frac{15}{4}$$

$$x = \frac{15}{4} \quad \text{»» Answer}$$

Problem 39) $\frac{2x}{5} - 3x = 10$

$$\frac{2x}{5} - 3x = 10$$

$$\frac{2x}{5} - 3x = 10$$

The rule for subtraction of fractions applies here:
$$\frac{A}{B} - \frac{C}{D} = \frac{A \cdot D - B \cdot C}{B \cdot D}$$

$$\frac{2x}{5} - \frac{3x}{1} = \frac{2x \cdot 1 - 5 \cdot 3x}{5 \cdot 1}$$
$$= \frac{2x - 15x}{5} = -\frac{13x}{5}$$

Combine the two elements with x

$$\frac{13x}{5} = 10$$

Now we have a nice and easy problem involving just one single x

$$-\frac{13x}{5} = 10$$

Moves as multiplication

Number in red is the operation performed in the current step
Number in blue is the operation performed in the previous step

36

$$-13x = 10 \cdot 5$$

$$x = \frac{50}{-13}$$

$$x = -\frac{50}{13} \text{ »» } \textbf{\textit{Answer}}$$

Problem 40) $\frac{x}{2} + \frac{x}{5} = 14$

$$\frac{x}{2} + \frac{x}{5} = 14$$

$$\frac{x}{2} + \frac{x}{5} = 14$$

Combine the two elements with x

The rule for addition of fractions applies here:

$$\frac{A}{B} + \frac{C}{D} = \frac{A \cdot D + B \cdot C}{B \cdot D}$$

$$\frac{x}{2} + \frac{x}{5} = \frac{2 \cdot x + x \cdot 5}{2 \cdot 5} = \frac{2x + 5x}{10} = \frac{7x}{10}$$

$$\frac{7x}{10} = 14$$

$$\frac{7x}{10} = 14$$

Moves as multiplication

$$7x = 14 \cdot 10$$

Moves as division

$$x = \frac{140}{7}$$

$$x = 20 \text{ »» } \textbf{\textit{Answer}}$$

Number in red is the operation performed in the current step

Number in blue is the operation performed in the previous step

Problem 41) $\frac{4x}{3} - \frac{3x}{4} = 4$

$$\frac{4x}{3} - \frac{3x}{4} = 4$$

$$\frac{4x}{3} - \frac{3x}{4} = 4$$

Combine the two elements with x

The rule for addition of fractions applies here:

$$\frac{A}{B} + \frac{C}{D} = \frac{A \cdot D + B \cdot C}{B \cdot D}$$

$$\frac{4x}{3} - \frac{3x}{4} = \frac{4x \cdot 4 - 3 \cdot 3x}{4 \cdot 3} = \frac{16x - 9x}{12} = \frac{7x}{12}$$

$$\frac{7x}{12} = 4$$

$$\frac{7x}{12} = 4$$

Moves as multiplication

$$7x = 4 \cdot 12$$

Moves as division

$$x = \frac{48}{7}$$

$$x = \frac{48}{7} \quad \text{»» } \textbf{\textit{Answer}}$$

Number in red is the operation performed in the current step
Number in blue is the operation performed in the previous step

Problem 42) $3x^3 - \frac{5x^3}{2} + 10 = -8$

$$3x^3 - \frac{5x^3}{2} + 10 = -8$$

$$3x^3 - \frac{5x^3}{2} + 10 = -8$$

> Combine the two elements with x

The rule for addition of fractions applies here:

$$\frac{A}{B} + \frac{C}{D} = \frac{A \cdot D + B \cdot C}{B \cdot D}$$

$$\frac{3x^3}{1} - \frac{5x^3}{2}$$
$$= \frac{3x^3 \cdot 2 - 1 \cdot 5x^3}{2}$$
$$= \frac{6x^3 - 5x^3}{2} = \frac{x^3}{2}$$

$$\frac{x^3}{2} + 10 = -8$$

$$\frac{x^3}{2} + 10 = -8$$

> Moves as negative

$$\frac{x^3}{2} = -8 - 10$$

> Moves as multiplication

$$x^3 = -18 \cdot 2$$

> Moves as a root

$$x = \sqrt[3]{-36}$$

$$x = \sqrt[3]{-36} \;\text{»» } \textbf{\textit{Answer}}$$

Number in red is the operation performed in the current step
Number in blue is the operation performed in the previous step

Problem 43) $6x^5 + \frac{x^5}{3} - 8 = 9$

$$6x^5 + \frac{x^5}{3} - 8 = 9$$

$$6x^5 + \frac{x^5}{3} - 8 = 9$$

Combine the two elements with x

$$\frac{19x^5}{3} - 8 = 9$$

$$\frac{19x^5}{3} - 8 = 9$$

Moves as positive

$$\frac{19x^5}{3} = 9 + 8$$

Moves as multiplication

$$19x^5 = 17 \cdot 3$$

Moves as division

$$x^5 = \frac{51}{19}$$

Moves as a root

$$x = \sqrt[5]{\frac{51}{19}}$$

$$x = \sqrt[5]{\frac{51}{19}} \quad \text{»» } \textit{Answer}$$

Number in red is the operation performed in the current step
Number in blue is the operation performed in the previous step

Problem 44) $3x^3 - \frac{5x^3}{2} + 10 = -8$

$$3x + 10 = \frac{x}{4} + 32$$

$$3x + 10 = \frac{x}{4} + 32$$

Moves as negative

The rule for addition of fractions applies here:

$$\frac{A}{B} + \frac{C}{D} = \frac{A \cdot D + B \cdot C}{B \cdot D}$$

$$\frac{3x}{1} - \frac{x}{4} = \frac{3x \cdot 4 - 1 \cdot x}{1 \cdot 4} =$$
$$\frac{12x - 1x}{4} = \frac{11x}{4}$$

$$3x - \frac{x}{4} + 10 = 32$$

$$3x - \frac{x}{4} + 10 = 32$$

Combine the two elements with x

Even when we have fractions, we have to group all x's together

$$\frac{11x}{4} + 10 = 32$$

$$\frac{11x}{4} + 10 = 32$$

Moves as negative

$$\frac{11x}{4} = 32 - 10$$

Moves as multiplication

$$11x = 22 \cdot 4$$

Moves as division

$$x = \frac{88}{11}$$

$$x = 8 \text{ »» } \textbf{\textit{Answer}}$$

Number in red is the operation performed in the current step
Number in blue is the operation performed in the previous step

41

<u>Fractions: practice problems</u>

1) $\dfrac{2}{1-x} = \dfrac{1}{x+1}$ 2) $\dfrac{10}{3x+5} = -\dfrac{4}{2x-1}$

3) $\dfrac{7}{9-x^2} = \dfrac{8}{2x^2-7}$ 4) $\dfrac{x}{2} + 8 - \dfrac{x}{3} + 7 = 16$

5) $\dfrac{3x}{5} - 2 = \dfrac{4x}{7} + 1$ 6) $2\left(\dfrac{5x}{6} + \dfrac{5}{3}\right) = -\left(\dfrac{3}{5} - \dfrac{4x}{3}\right)$

Number in red is the operation performed in the current step
Number in blue is the operation performed in the previous step

More parenthesis

Yes, we are coming back to parenthesis, just when you thought they were over unfortunately. Actually, if we look back at the fraction problems, many times we ended up with a parenthesis situation.

Now we are mixing parenthesis with other operations learned in the past. The idea remains; first distribute the parenthesis, then group all the x's in the same side and finally, solve for x.

Problem 45) $5(2x - 3) + 2x = 3x + 21$

$$5(2x - 3) + 2x = 3x + 21$$

$$5(2x - 3) + 2x = 3x + 21$$

Distribute the 5 inside of the parenthesis

$$5 \cdot 2x - 5 \cdot 3 + 2x = 3x + 21$$

$$10x - 15 + 2x = 3x + 21$$

$$10x - 15 + 2x = 3x + 21$$

First let's take care of the parenthesis

And now, as you probably guessed, we group them together

Moves as negative

$$10x + 2x - 3x - 15 = 21$$

$$10x + 2x - 3x - 15 = 21$$

Combine all the elements with x

There we go, a problem we are more used to solve

Just in case so many x's made you lose track::
$$10x + 2x - 3x =$$
$$x(10 + 2 - 3) =$$
$$9x$$

$$9x - 15 = 21$$

$$9x - 15 = 21$$

Moves as positive

$$9x = 21 + 15$$

Moves as division

$$x = \frac{36}{9}$$

$$x = 4 \text{ »» } \textbf{Answer}$$

Number in red is the operation performed in the current step
Number in blue is the operation performed in the previous step

Problem 46) $3(4 - 6x) + 8x + 10 = -2x + 30$

$$3(4 - 6x) + 8x + 10 = -2x + 30$$

$$3(4 - 6x) + 8x + 10 = -2x + 30$$

And now, as you probably guessed, we group them together

Distribute the 5 inside of the parenthesis

First let's take care of the parenthesis

$$3 \cdot 4 - 3 \cdot 6x + 8x + 10 = -2x + 30$$

$$12 - 18x + 8x + 10 = -2x + 3$$

$$12 - 18x + 8x + 10 = -2x + 30$$

Moves as positive

$$-18x + 8x + 2x + 12 + 10 = 30$$

$$-18x + 8x + 2x + 12 + 10 = 30$$

Combine all the elements with x

There we go, a problem we are more used to solve

Just in case so many x's made you lose track:
$-18x + 8x + 2x =$
$x(-18 + 8 + 2) =$
$-8x$

$$-8x + 12 + 10 = 30$$

$$-8x + 12 + 10 = 30$$

Moves as negative

$$-8x = 30 - 12 - 10$$

Moves as division

$$x = \frac{8}{-8}$$

$$x = -1 \text{ »» } \textbf{Answer}$$

Just a friendly reminder about only one element of the fraction (denominator in this case) having a negative, therefore the answer is negative

Number in red is the operation performed in the current step
Number in blue is the operation performed in the previous step

44

Problem 47) $4(2x - 5) + 6x = 7(x - 5) + 2x$

$$4(2x - 5) + 6x = 7(x - 5) + 2x$$

$$4(2x - 5) + 6x = 7(x - 5) + 2x$$

And now, as you probably guessed, we group them together

Distribute the 4 and 7 inside of the parenthesis

First let's take care of the parenthesis

$$4 \cdot 2x - 4 \cdot 5 + 6x = 7 \cdot x - 7 \cdot 5 + 2x$$

$$8x - 20 + 6x = 7x - 35 + 2x$$

$$8x - 20 + 6x = 7x - 35 + 2x$$

Moves as negative

$$8x + 6x - 7x - 2x - 20 = -35$$

$$8x + 6x - 7x - 2x - 20 = -35$$

Combine all the elements with x

There we go, a problem we are more used to solve

$$5x - 20 = -35$$

$$5x - 20 = -35$$

Just in case so many x's made you lose track:
$$8x + 6x - 7x - 2x$$
$$= x(8 + 6 - 7 - 2)$$
$$= 5x$$

Moves as positive

$$5x = -35 + 20$$

Moves as division

$$x = \frac{-15}{5}$$

$$x = -3 \text{ »» } \textbf{\textit{Answer}}$$

Just a friendly reminder about only one element of the fraction (numerator in this case) having a negative, therefore the answer is negative

Problem 48) $12(x^2 - 6) + 4x^2 = 10x^2 + 8$

$$12(x^2 - 6) + 4x^2 = 10x^2 + 8$$

$$12(x^2 - 6) + 4x^2 = 10x^2 + 8$$

First let's take care of the parenthesis

Distribute the 12 inside of the parenthesis

$$12 \cdot x^2 - 12 \cdot 6 + 4x^2 = 10x^2 + 8$$

$$12x^2 - 72 + 4x^2 = 10x^2 + 8$$

$$12x^2 - 72 + 4x^2 = 10x^2 + 8$$

And now, as you probably guessed, we group them together

Moves as negative

$$12x^2 + 4x^2 - 10x^2 - 72 = 8$$

$$12x^2 + 4x^2 - 10x^2 - 72 = 8$$

There we go, a problem we are more used to solve

Combine all the elements with x

Just in case so many x's made you lose track:
$$12x^2 + 4x^2 - 10x^2$$
$$= x^2(12 + 4 - 10)$$
$$= 6x^2$$

$$6x^2 - 72 = 8$$

$$6x^2 - 72 = 8$$

Moves as positive

$$6x^2 = 8 + 72$$

Moves as division

$$x^2 = \frac{80}{6}$$

Moves as a root

Don't forget to simplify the fraction inside of the root:

$$\frac{80 \div 2}{6 \div 2} = \frac{40}{3}$$

$$x = \sqrt[2]{\frac{80}{6}}$$

$$x = \sqrt{\frac{40}{3}} \ \text{ and } \ x = -\sqrt{\frac{40}{3}} \ \text{»» \textit{Answers}}$$

Number in red is the operation performed in the current step
Number in blue is the operation performed in the previous step

Problem 49) $7(6 - x^2) + 25 = 8(-10 - 3x^2) + 8x^2$

$$7(6 - x^2) + 25 = 8(-10 - 3x^2) + 8x^2$$

$$7(6 - x^2) + 25 = 8(-10 - 3x^2) + 8x^2$$

First let's take care of the parenthesis

Distribute the 7 and 8 inside of the parenthesis

$$7 \cdot 6 - 7 \cdot x^2 + 25 = 8 \cdot - 10 - 8 \cdot 3x^2 + 8x^2$$

$$42 - 7x^2 + 25 = -80 - 24x^2 + 8x^2$$

Moves as positive

$$42 - 7x^2 + 25 = -80 - 24x^2 + 8x^2$$

Moves as negative

And now, as you probably guessed, we group them together

There we go, a problem we are more used to solve

$$-7x^2 + 24x^2 - 8x^2 + 42 + 25 = -80$$

$$-7x^2 + 24x^2 - 8x^2 + 42 + 25 = -80$$

Combine all the elements with x

$$9x^2 + 42 + 25 = -80$$

$$9x^2 + 42 + 25 = -80$$

Just in case so many x's made you lose track:
$$-7x^2 + 24x^2 - 8x^2$$
$$= x^2(-7 + 24 - 8)$$
$$= 9x^2$$

Moves as negative

$$9x^2 = -80 - 42 - 25$$

Moves as division

$$x^2 = \frac{-147}{9}$$

Yes, we repeated all the notifications looking like this square for all the problems in this section... but we just did not want you to get lost

Moves as a root

$$x = \sqrt[2]{-147/9}$$

The square root of a negative number does NOT exist, trust us!

$$x = \sqrt{-147/9} \gg\gg \textbf{\textit{No Answer}}!!$$

Number in red is the operation performed in the current step
Number in blue is the operation performed in the previous step

Chapter three:
Different x's
and different powers

Common factor and zero product property

In the previous chapter we started evaluating problems having more than one x, as long as all the x's had the same exponent we were able to just add them or subtract them together. Unfortunately, math likes to escalate the difficulty of its problems continuously, therefore, once we finally mastered the problems containing more than one x, now we have to face problems involving more than one x and more than one exponent.

In order to solve these problems, we will need to use some basic algebra rules in order to turn a multiple x and multiple exponent problem into an easier problem. Sometimes it is really easy to forget some of them so we will write them for you:

Common factor: Anytime there is a series of additions/subtractions next to each other, and they are all being multiplied by exactly the same number or letter, we can take that repeated number/letter out. Let's view some examples to understand this rule better

Example 1:

$$3x + 3 = 10$$

The two elements on the left side contain a 3, so we can take a common factor of 3

$$3x + 3 = 10$$

First we take the 3 out from each element
$$3(x + 1) = 10$$

Then we write it next to a parenthesis multiplying the remaining numbers

Note how the numbers inside of the parenthesis have changed. Remember we just took a 3 from them, is like a division

$$3x + 3 \rightarrow by\ taking\ the\ 3\ out \rightarrow 3\left(\frac{3x}{3} + \frac{3}{3}\right) \rightarrow 3(x + 1)$$

So our final simplified version would be:
$$3(x + 1) = 10$$

But let's do a few more examples in order to keep this concept 100% clear

Example 2: $4x - 4 = 5$

In this case the 4 is repeated in both elements on the left, let's take it out

$$4x - 4 \rightarrow take\ out\ the\ 4 \rightarrow 4\left(\frac{4x}{4} - \frac{4}{4}\right) = 4(x - 1)$$

Number in red is the operation performed in the current step
Number in blue is the operation performed in the previous step

So our final simplified version becomes:

$$4(x - 1) = 5$$

Sometimes the common factor is not as straight forward and we may have to think harder to find it, like in these cases:

Example 3: $3x - 9 = 8$

At plain sight there is no repeated number, but if we think carefully, both 3 and 9 are divisible by 3, so we can take a 3 as a common factor

$$3x - 9 = 8 \rightarrow take\ a\ 3\ out \rightarrow 3\left(\frac{3x}{3} - \frac{9}{3}\right) \rightarrow 3(x - 3)$$

Our simplified version:

$$3(x - 3) = 8$$

As long as all numbers are divisible by the same number, we can take a common factor, here is another example.

Example 4: $16x + 24 = 2$

Even though this one is harder to see, both are divisible by 4

$$16x + 24 \rightarrow take\ out\ a\ 4 \rightarrow 4\left(\frac{16x}{4} + \frac{24}{4}\right) = 4(4x + 6)$$

Final version:

$$4(4x + 6) = 2$$

But taking a common factor works even if we have more than just 2 elements. We have not learned how to solve problems with different letters yet, but for the purpose of explanation we will use one of those problems for the next example.

Example 5: $4x - 10y + 8z = 7$

All the numbers on the left side of the equation are divisible by 2

$$4x - 10y + 8z \rightarrow take\ out\ a\ 2 \rightarrow 2\left(\frac{4x}{2} - \frac{10y}{2} + \frac{8z}{2}\right) \rightarrow 2(2x - 5y + 4z)$$

Final simplified version:

$$2(2x - 5y + 4z) = 7$$

Number in red is the operation performed in the current step
Number in blue is the operation performed in the previous step

49

As seen, it doesn't matter how many elements there are, as long as they are all divisible by the same value, a common factor can be taken. Now let's not be closed minded, the common factor doesn't necessary has to be a number, it could be   a variable! Yes, a variable.

Example 6: $4x^2 - 5x^3 - 3x^4 = 5$

Even though the numbers in the elements do not have anything in common, their variables do.

They all contain at least one x, meaning we can take a common factor of x.

The rule is, we take out the smallest x in the park. In this case x²

$$4x^2 - 5x^3 - 3x^4 \rightarrow take\ out\ an\ x^2 \rightarrow x^2\left(\frac{4x^2}{x^2} - \frac{5x^3}{x^2} - \frac{3x^4}{x^2}\right) = x^2(4 - 5x - 3x^2)$$

Therefore, our final simplified version:

$$x^2\left(4 - 5x - 3x^2\right) = 5$$

And there will be some cases in which we may have to take a variable AND a number as common factors.

Example 7: $6x^5 - 27x + 15x^3 = -10$

All elements have numbers divisible by 3

All elements have an x; we can take out the smallest; just x for this case

$$6x^5 - 27x + 15x^3 \rightarrow take\ out\ a\ 3\ and\ an\ x \rightarrow 3x\left(\frac{6x^5}{3x} - \frac{27x}{3x} + \frac{15x^3}{3x}\right) \rightarrow 3x(2x^4 - 9 + 5x^2)$$

Our final simplified version is:

$$3x\left(2x^4 - 9 + 5x^2\right) = -10$$

An important information to keep in mind for common factors is: every single element in one side of the equal sign MUST be divisible by the same number or letter, it only takes one element to not have that divisor for the common factor rule to fail. Check this:

Example 8: $5x^2 + 15x^4 - 20x + 35x^3 - 8 = 60$

There are five elements in the left side of the equation

Four of them have x and the same four are divisible by 5. The last element (-8) does not have an x and is not divisible by 5. Common factor can NOT be taken.

Hint: We could move the -8 to the right side of the equation and then take a common factor

Number in red is the operation performed in the current step
Number in blue is the operation performed in the previous step

Zero product property: Next important rule we must remember for this chapter is this one. The zero product property tells us that every time we have two or more elements multiplying each other and they are equal to 0, we can write them as different equations equal to 0. Then again, let's view some examples to clarify the idea:

Example 1: $x(x + 1) = 0$

The x on the far left is multiplying the operation inside of the parenthesis and they are equal to 0

They can be written as separate operations:

$$x = 0 \quad and \quad x + 1 = 0$$

Once the **zero product property** has been applied, we can solve the problem as two different equations. Check some more examples:

Example 2: $(x - 2)(x^2 - 5) = 0$

Both parenthesis are multiplying each other and the other side of the equal sign has a 0

The zero product property can be applied here:

$$x - 2 = 0 \quad and \quad x^2 - 5 = 0$$

But this property works even if we have more than two elements. As long as they are multiplying each other, it applies.

Example 3: $4x^3(x + 4)(10 - x^3)\sqrt{2x} = 0$

Since all elements are multiplying each other:

$$4x^3 = 0 \; ; \; x + 4 = 0 \; ; \; 10 - x^3 = 0 \quad and \quad \sqrt{2x} = 0$$

Yes, we could also write 4 = 0 since the four is a multiplication but that makes absolutely no sense

Which is why we keep it next to the x³, trust us, it won't make any difference.

Number in red is the operation performed in the current step
Number in blue is the operation performed in the previous step

51

Problem 50) $x^2 + x = 0$

$$x^2 + x = 0$$

$x^2 + x = 0$

Take a common factor of x

$x(x+1) = 0$

$x(x+1) = 0$

Apply the zero product property

Once we apply the zero product property, is just a matter of solving the remaining equations

Don't forget to divide by the common factor:

$$x\left(\frac{x^2}{x} + \frac{x}{x}\right) = x(x+1)$$

$x = 0$

$x = 0$

$x + 1 = 0$

$x + 1 = 0$

Moves as negative

$x = 0 - 1$

$x = -1$

$x = 0 \ and \ x = -1 \ »» \ \textbf{Answers}$

Problem 51) $3x^2 - 9x = 0$

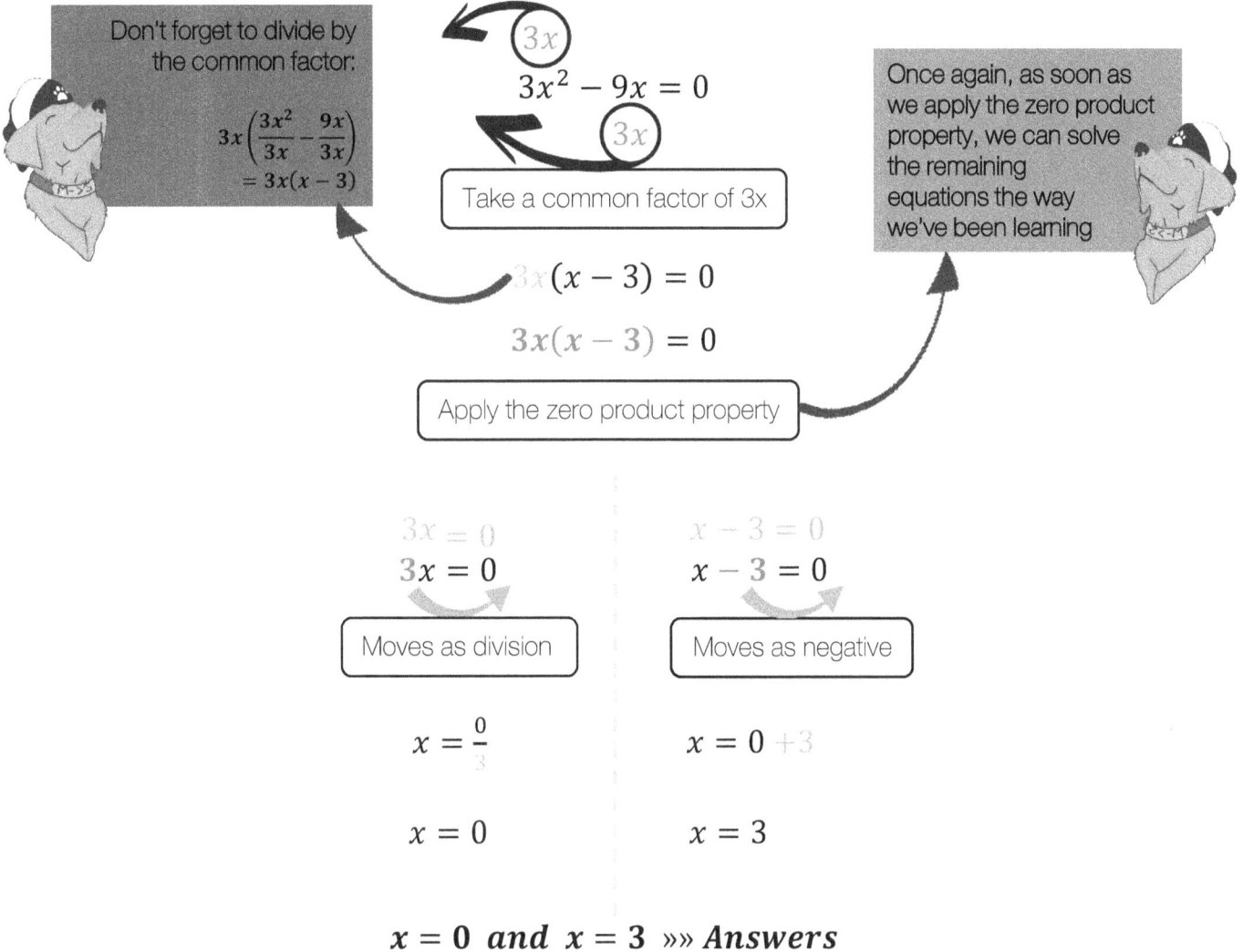

$$3x^2 - 9x = 0$$

$3x$

$$3x^2 - 9x = 0$$

$3x$

Don't forget to divide by the common factor:

$$3x\left(\frac{3x^2}{3x} - \frac{9x}{3x}\right)$$
$$= 3x(x-3)$$

Take a common factor of 3x

Once again, as soon as we apply the zero product property, we can solve the remaining equations the way we've been learning

$$3x(x-3) = 0$$

$$3x(x-3) = 0$$

Apply the zero product property

$3x = 0$

$$3x = 0$$

$x - 3 = 0$

$$x - 3 = 0$$

Moves as division

Moves as negative

$$x = \frac{0}{3}$$

$$x = 0 + 3$$

$$x = 0$$

$$x = 3$$

$$x = 0 \text{ and } x = 3 \text{ »» Answers}$$

Number in red is the operation performed in the current step
Number in blue is the operation performed in the previous step

53

Problem 52) $16x^3 - 6x = 0$

$$16x^3 - 6x = 0$$

$(2x)$

$$16x^3 - 6x = 0$$

$(2x)$

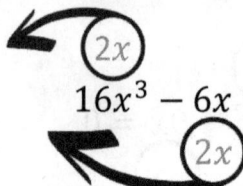

Take a common factor of 2x

Don't forget to divide by the common factor:

$$2x\left(\frac{16x^3}{2x} - \frac{6x}{2x}\right) =$$

$$2x(8x^2 - 3)$$

Once again, as soon as the we apply the zero product property, we can solve the remaining equations the way we've been learning

$$2x(8x^2 - 3) = 0$$

$$2x(8x^2 - 3) = 0$$

Apply the zero product property

$2x = 0$ | $8x^2 - 3 = 0$

$$2x = 0$$

Moves as division

$$8x^2 - 3 = 0$$

Moves as positive

$$x = \frac{0}{2}$$

$$8x^2 = 0 + 3$$

$$x = 0$$

Moves as division

$$x^2 = \frac{3}{8}$$

Moves as a root

$$x = \sqrt[2]{\frac{3}{8}}$$

$$x = \pm\sqrt{\frac{3}{8}}$$

$$x = 0; x = \sqrt{\frac{3}{8}} \text{ and } x = -\sqrt{\frac{3}{8}} \text{ »» Answers}$$

Number in red is the operation performed in the current step
Number in blue is the operation performed in the previous step

54

Problem 53) $5x^3 + 25x = 0$

$$5x^3 + 25x = 0$$

$(5x)$

$$5x^3 + 25x = 0$$

$(5x)$

Take a common factor of 5x

Don't forget to divide by the common factor:

$$5x\left(\frac{5x^3}{5x} - \frac{25x}{5x}\right) =$$

$$5x(x^2 + 5)$$

$5x(x^2 + 5) = 0$

$$5x(x^2 + 5) = 0$$

Apply the zero product property

$5x = 0$ | $x^2 + 5 = 0$

$$5x = 0$$

Moves as division

$$x^2 + 5 = 0$$

Moves as negative

$$x = \frac{0}{5}$$

$$x = 0$$

$$x^2 = 0 - 5$$

Moves as a root

$$x = \sqrt[2]{-5}$$

There is no such a thing as the square root of a negative number, it simply does NOT exist, therefore this side has NO SOLUTION!

$$x = 0 \text{ »» } \textbf{\textit{Answer}}$$

Number in red is the operation performed in the current step
Number in blue is the operation performed in the previous step

55

Problem 54) $2x^4 - 8x^2 = 0$

$$2x^4 - 8x^2 = 0$$

Don't forget to divide by the common factor:

$$2x^2\left(\frac{2x^4}{2x^2} - \frac{8x^2}{2x^2}\right) =$$
$$2x^2(x^2 - 4)$$

$2x^2$

$$2x^4 - 8x^2 = 0$$

$2x^2$

Take a common factor of 2x²

$$2x^2(x^2 - 4) = 0$$

$$2x^2(x^2 - 4) = 0$$

Apply the zero product property

$2x^2 = 0$	$x^2 - 4 = 0$
$2x^2 = 0$	$x^2 - 4 = 0$
Moves as division	Moves as positive
$x^2 = \dfrac{0}{2}$	$x^2 = 0 + 4$
Moves as a root	Moves as a root
$x = \sqrt[2]{0}$	$x = \sqrt[2]{4}$
$x = 0$	$x = \pm 2$

$$x = 0; x = 2 \text{ and } x = -2 \text{ »» } \textbf{\textit{Answer}}$$

Number in red is the operation performed in the current step
Number in blue is the operation performed in the previous step

Problem 55) $3x^2 + 5x = 2x^2 - 3x$

$$3x^2 + 5x = 2x^2 - 3x$$

Moves as negative

$$3x^2 + 5x = \;2x^2 - 3x$$

Moves as positive

$$3x^2 - 2x^2 + 5x + 3x = 0$$

$$3x^2 - 2x^2 + 5x + 3x = 0$$

$$x^2 + 8x = 0$$

BOO!

That idea about grouping all x's on the same side of the equation will haunt us for the rest of our math life.

Don't forget to divide by the common factor:

$$x\left(\frac{x^2}{x} + \frac{8}{x}\right) = x(x+8)$$

$$x^2 + 8x = 0$$

Take a common factor of x

$$x(x+8) = 0$$

$$x(x+8) = 0$$

Apply the zero product property

$x = 0$	$x + 8 = 0$
$x = 0$	$x + 8 = 0$
	Moves as negative
	$x = -8$
	$x = -8$

$$x = 0 \; and \; x = -8 \;\text{»» } Answers$$

Number in red is the operation performed in the current step
Number in blue is the operation performed in the previous step

Problem 56) $6x^5 + 13x^2 = 5x^5 - 14x^2$

$$6x^5 + 13x^2 = 5x^5 - 14x^2$$

Moves as negative

$$6x^5 + 13x^2 = 5x^5 - 14x^2$$

Moves as positive

$$6x^5 - 5x^5 + 13x^2 + 14x^2 = 0$$

$$6x^5 - 5x^5 + 13x^2 + 14x^2 = 0$$

$$x^5 + 27x^2 = 0$$

Grouping all x's in one side of the equal sign will always help us solve the problem

Don't forget to divide by the common factor:

$$x^2 \left(\frac{x^5}{x^2} + \frac{27x^2}{x^2} \right) = x^2(x^3 + 27)$$

x^2

$$x^5 + 27x^2 = 0$$

x^2

Take a common factor of x²

$$x^2(x^3 + 27) = 0$$

$$x^2\left(x^3 + 27\right) = 0$$

Apply the zero product property

$$x^2 = 0 \qquad\qquad x^3 + 27 = 0$$

$$x^3 + 27 = 0$$

$$x^2 = 0$$

Moves as a root

Moves as negative

$$x = \sqrt[2]{0}$$

$$x^3 = -27$$

$$x = 0$$

Moves as a root

$$x = \sqrt[3]{-27}$$

$$x = -3$$

$$x = 0 \text{ and } x = -3 \text{ »» } \textbf{Answers}$$

Number in red is the operation performed in the current step
Number in blue is the operation performed in the previous step

Problem 57) $6x^4 - 15x^2 = 4x^4 + 17x^2$

$$6x^4 - 15x^2 = 4x^4 + 17x^2$$

$$6x^4 - 15x^2 = 4x^4 + 17x^2$$

Moves as negative

$$6x^4 - 4x^4 - 15x^2 - 17x^2 = 0$$

$$6x^4 - 4x^4 - 15x^2 - 17x^2 = 0$$

$$2x^4 - 32x^2 = 0$$

Grouping all x's in one side of the equal sign will always help us solve the problem

Don't forget to divide by the common factor:

$$2x^2\left(\frac{2x^4}{2x^2} - \frac{32x^2}{2x^2}\right) = 2x^2(x^2 - 16)$$

$$2x^4 - 32x^2 = 0$$

x^2

$2x^2$

Take a common factor of $2x^2$

$$2x^2(x^2 - 16) = 0$$

$$2x^2(x^2 - 16) = 0$$

Apply the zero product property

$2x^2 = 0$	$x^2 - 16 = 0$
$2x^2 = 0$	$x^2 - 16 = 0$
Moves as division	Moves as positive
$x^2 = \dfrac{0}{2}$	$x^2 = 0 + 16$
Moves as a root	Moves as a root
$x = \sqrt[2]{0}$	$x = \sqrt[2]{16}$
$x = 0$	$x = \pm 4$

$$x = 0; x = 4 \text{ and } x = -4 \text{ »» } Answers$$

Number in red is the operation performed in the current step
Number in blue is the operation performed in the previous step

Common factor and zero product property: practice problems

1) $4x^2 + 6x = 0$ 2) $3x^4 - 12x^2 = 0$

3) $7x + 8x^3 = 4(3x + x^3)$ 4) $3(2x^2 - 5x) = 2(x^2 - 20x)$

5) $\dfrac{2x}{3} - \dfrac{4x^2}{9} = 0$ 6) $\dfrac{x}{2} + \dfrac{x^2}{5} = x + x^2$

Quadratic equations: factorization

Quadratic equations are part of math's fancy repertory. Just like English, when we start using words such as therefore, hence, nonetheless, consequently or simply speak in British accent to appear more distinguished, quadratic equations gives us the opportunity to look more proficient in math once we start using them. But first what is first, we can't pretend to brag about quadratic equations if we don't know what they are.

In theory, a quadratic equation is any equation whose variable with higher exponent is a 2. So any of the examples bellow is a quadratic equation:

$$a)\ 0 = 3x^2 + x \quad b)\ 4 = -x^2 - \frac{2}{7} \quad c)\ 60 = \frac{x^2}{2} - 5x + 10 \quad d)\ 4 = x^2$$

None of those equation has an x with exponent higher than 2, therefore, they are all quadratic equations. Except for example c, we can solve them all by using the conventional methods we've learned through the book. For purposes of simplification and making explanations more understandable, we are only going to call quadratic equations to those fulfilling the generic quadratic equation:

<u>Generic quadratic equation:</u>
$$Ax^2 + Bx + C = 0$$

Where A, B and C represent any number, either positive or negative

The generic quadratic equation consists of an element having an x squared, an element having a normal x, an element not having any x and being equal to 0. As soon as any of these is missing, we can solve for x using all the other methods learned before, which is why we will call quadratic equation only to those having all of the components. Check these examples of full quadratic equations:

$$a)\ x^2 - 3x + 8 = 0 \quad b)\ 0 = 4 - 5x + 10x^2 \quad c)\ x - 7x^2 + 100 = 0$$

Quadratic equations require special solving methods, two of them will be explained in this book. Let's learn first factorization. For factorization we will turn the function into a factor multiplication and then apply the zero product property to get two separate answers:

$$Ax^2 + Bx + C = 0$$

After factorization
$$(x - Factor1)(x - Factor2) = 0$$

After zero product property

$$x - factor1 = 0 \quad and \quad x - factor2 = 0$$

The process will be very detailed explained in the following examples. It is the same process every time.

Number in red is the operation performed in the current step
Number in blue is the operation performed in the previous step

61

Problem 58) $x^2 + 5x - 6 = 0$

$$x^2 + 5x - 6 = 0$$

$$1x^2 + 5x - 6 = 0$$

$$A = 1 \; ; B = 5 \; ; C = -6$$

- A is the number multiplying the x^2. If you don't see any, assume a 1
- B is the number multiplying the normal x. If you don't see any, assume a 1
- C is the number without x

Identify the values for A, B and C

$$A \cdot C = 1 \cdot -6 = -6$$

Multiply A and C together.

Usually takes a bit of time guessing and checking, but like in cooking programs, we are giving you the final result right away

Now we must find two numbers that meet:

- When added each other are equal to B (5 in this problem)
$$6 - 1 = 5$$
- When multiplied each other are equal to A·C (-6 in this problem)
$$6 \cdot (-1) = -6$$

So our magic numbers are **positive 6** and **negative 1**

$$x^2 + 5x - 6 = 0$$

Split this number into the two magic numbers

$$x^2 - 1x + 6x - 6 = 0$$

$$\{x^2 - 1x\}\{+6x - 6\} = 0$$

A reminder about the common factor process
$$x\left(\frac{x^2}{x} - \frac{1x}{x}\right) = x(x - 1)$$

Separate the new equation in two groups

A reminder about the common factor process
$$6\left(\frac{6x}{6} - \frac{6}{6}\right) = 6(x - 1)$$

$$\{x^2 - 1x\}\{+6x - 6\} = 0$$
x

6

Take a common factor of x from the green (light) group and a 6 from the purple (dark) group

$$x(x - 1) + 6(x - 1) = 0$$

$$x(x - 1) + 6(x - 1) = 0$$

Equal values inside of the parenthesis, this is a good sign!!

Let's evaluate what we have:

- Equal value inside of the parenthesis, in green = (x - 1)
- Leftovers outside of the parenthesis, in purple = (x + 6)
- New equation resulting from multiplying both of the above: $(x - 1)(x + 6) = 0$

Number in red is the operation performed in the current step
Number in blue is the operation performed in the previous step

We did it! The equation is now on factor form

The main idea of the factorization process is to have equal values inside of the parenthesis

Then have factor form by multiplying the repeated parenthesis by the leftovers outside of the parenthesis.

$$(x - 1)(x + 6) = 0$$

Apply the zero product property

$x - 1 = 0$ $\qquad\qquad$ $x + 6 = 0$

$x - 1 = 0$ $\qquad\qquad$ $x + 6 = 0$

Moves as positive $\qquad\qquad$ Moves as negative

$x = 0 + 1$ $\qquad\qquad$ $x = 0 - 6$

$x = 1$ $\qquad\qquad$ $x = -6$

$$x = 1 \ and \ x = -6 \text{ »» } Answers$$

Number in red is the operation performed in the current step
Number in blue is the operation performed in the previous step

63

Problem 59) $x^2 - 8x + 12 = 0$

$$x^2 - 8x + 12 = 0$$

$$1x^2 - 8x + 12 = 0$$

$$A = 1 \; ; B = -8 \; ; C = 12$$

Identify the values for A, B and C

- A is the number multiplying the x^2. If you don't see any, assume a 1
- B is the number multiplying the normal x. If you don't see any, assume a 1
- C is the number without x

$$A \cdot C = 1 \cdot 12 = 12$$

Multiply A and C together.

Even though we are giving you the magic numbers, try it on your own first. It is tricky at first

Now we must find two numbers that meet:

- When added each other are equal to B (-8 in this problem)

$$-6 - 2 = -8$$

- When multiplied each other are equal to A·C (-12 in this problem)

$$(-6) \cdot (-2) = 12$$

So our magic numbers are **negative 6** and **negative 2**

$$x^2 - 8x + 12 = 0$$

Split this number into the two magic numbers

$$x^2 - 2x - 6x + 12 = 0$$

$$\{x^2 - 2x\} \{-6x + 12\} = 0$$

Separate the new equation in two groups

A reminder about the common factor process

$$x\left(\frac{x^2}{x} - \frac{2x}{x}\right) =$$
$$x(x - 2)$$

$$\{x^2 - 2x\} \{-6x + 12\} = 0$$

A reminder about the common factor process

$$-6\left(\frac{-6x}{-6} - \frac{12}{-6}\right) =$$
$$-6(x - 2)$$

Take a common factor of x from the green (light) group and a -6 from the purple (dark) group

$$x(x - 2) - 6(x - 2) = 0$$

$$x(x - 2) - 6(x - 2) = 0$$

Equal values inside of the parenthesis, this is a good sign!!

Number in red is the operation performed in the current step
Number in blue is the operation performed in the previous step

Let's evaluate what we have:

- Equal value inside of the parenthesis, in green = (x - 2)
- Leftovers outside of the parenthesis, in purple = (x - 6)
- New equation resulting from multiplying both of the above: $(x - 2)(x - 6) = 0$

We did it! The equation is now on factor form

The main idea of the factorization process is to have equal values inside of the parenthesis

Then have factor form by multiplying the repeated parenthesis by the leftovers outside of the parenthesis.

$$(x - 2)(x - 6) = 0$$

Apply the zero product property

$$x - 2 = 0 \qquad\qquad x - 6 = 0$$

$$x - 2 = 0 \qquad\qquad x - 6 = 0$$

Moves as positive | Moves as positive

$$x = 0 + 2 \qquad\qquad x = 0 + 6$$

$$x = 2 \qquad\qquad x = 6$$

$$x = 2 \ and \ x = 6 \ »» \ Answers$$

Number in red is the operation performed in the current step
Number in blue is the operation performed in the previous step

65

Problem 60) $x^2 + 4x + 4 = 0$

$$x^2 + 4x + 4 = 0$$

$$1x^2 + 4x + 4 = 0$$

$$A = 1 \; ; B = 4 \; ; C = 4$$

Identify the values for A, B and C

- A is the number multiplying the x^2. If you don't see any, assume a 1
- B is the number multiplying the normal x. If you don't see any, assume a 1
- C is the number without x

$$A \cdot C = 1 \cdot 4 = 4$$

Multiply A and C together.

Trying to find these magic numbers is not easy, it requires practice!

Now we must find two numbers that meet:

- When added each other are equal to B (4 in this problem)

$$2 + 2 = 4$$

- When multiplied each other are equal to A·C (-6 in this problem)

$$(2) \cdot (2) = 2$$

So our magic numbers are **positive 2** and **positive 2**

$$x^2 + 4x + 4 = 0$$

Split this number into the two magic numbers

$$x^2 + 2x + 2x + 4 = 0$$

$$\{x^2 + 2x\}\{+2x + 4\} = 0$$

Separate the new equation in two groups

A reminder about the common factor process

$$x\left(\frac{x^2}{x} + \frac{2x}{x}\right) =$$
$$x(x + 2)$$

$$\text{(x)} \quad \{x^2 + 2x\}\{+2x + 4\} = 0 \quad \text{(2)}$$

A reminder about the common factor process

$$2\left(\frac{2x}{2} + \frac{4}{2}\right) =$$
$$2(x + 2)$$

Take a common factor of x from the green (light) group and a 2 from the purple (dark) group

$$x(x + 2) + 2(x + 2) = 0$$

$$x(x + 2) + 2(x + 2) = 0$$

Equal values inside of the parenthesis, this is a good sign!!

Number in red is the operation performed in the current step
Number in blue is the operation performed in the previous step

Let's evaluate what we have:

- Equal value inside of the parenthesis, in green = (x + 2)
- Leftovers outside of the parenthesis, in purple = (x + 2)
- New equation resulting from multiplying both of the above: $(x + 2)(x + 2) = 0$

We did it! The equation is now on factor form
The main idea of the factorization process is to have equal values inside of the parenthesis
Then have factor form by multiplying the repeated parenthesis by the leftovers outside of the parenthesis.

$$(x + 2)(x + 2) = 0$$

Apply the zero product property

$$x + 2 = 0 \qquad\qquad x + 2 = 0$$

$$x + 2 = 0 \qquad\qquad x + 2 = 0$$

Moves as negative $\qquad$ Moves as negative

$$x = 0 - 2 \qquad\qquad x = 0 - 2$$

$$x = -2 \qquad\qquad x = -2$$

$$x = -2 \quad \text{»» } \textbf{\textit{Answer}}$$

Quadratic equations: factorization: practice problems

1) $x^2 + 7x - 8 = 0$ $\qquad$ 2) $x^2 - 15x + 50 = 0$

3) $x^2 + 2x = 8x + 7$ $\qquad$ 4) $11x + 4 = -x^2 - 6$

5) $x^2 - 11x + 30 = 0$ $\qquad$ 6) $x^2 - 2x - 195$

Number in red is the operation performed in the current step
Number in blue is the operation performed in the previous step

Quadratic equations: quadratic formula

There is a second method used to solve for quadratic equations. First take a look at the generic quadratic equation:

$$Ax^2 + Bx + C = 0$$

Where A, B and C represent any number positive or negative

The numbers (A, B and C) multiplying each element are called coefficients, they are allowed to be any number, either positive or negative. Once we have identified each coefficient we can plug them into the following formula to solve for x:

$$x = \frac{-B \pm \sqrt{B^2 - 4 \cdot A \cdot C}}{2 \cdot A}$$

The formula itself will provide us with the two answers of x. Let's get some practice with this method.

Problem 61) $x^2 + 5x - 6 = 0$

$$x^2 + 5x - 6 = 0$$

$$1x^2 + 5x - 6 = 0$$

$$A = 1 \; ; B = 5 \; ; C = -6$$

Identify the values for A, B and C

$$\frac{-B \pm \sqrt{B^2 - 4 \cdot A \cdot C}}{2 \cdot A} \rightarrow \frac{-(5) \pm \sqrt{(5)^2 - 4 \cdot (1) \cdot (-6)}}{2 \cdot (1)}$$

Plug the values into the formula

$$= \frac{-(5) \pm \sqrt{(5)^2 - 4 \cdot (1) \cdot (-6)}}{2 \cdot (1)}$$

$$= \frac{-5 \pm \sqrt{25 + 24}}{2 \cdot (1)} = \frac{-5 \pm \sqrt{49}}{2 \cdot (1)}$$

Start with the inside of the root

$$= \frac{-5 \pm \sqrt{49}}{2 \cdot (1)}$$

$$= \frac{-5 \pm 7}{2}$$

Solve the root and the multiplication in the denominator

Number in red is the operation performed in the current step
Number in blue is the operation performed in the previous step

$$= \frac{-5 \pm 7}{2}$$

Split the plus or minus sign to get the two answers

$$x = \frac{-5 + 7}{2} \qquad\qquad x = \frac{-5 - 7}{2}$$

$$x = \frac{2}{2} \qquad\qquad x = \frac{-12}{2}$$

$$x = 1 \qquad\qquad x = -6$$

$x = 1 \; and \; x = -6$ »» *Answers*

Number in red is the operation performed in the current step
Number in blue is the operation performed in the previous step

69

Problem 62) $x^2 - 8x + 12 = 0$

$$x^2 - 8x + 12 = 0$$

$$1x^2 - 8x + 12 = 0$$

$$A = 1 \; ; B = -8 \; ; C = 12$$

Identify the values for A, B and C

$$\frac{-B \pm \sqrt{B^2 - 4 \cdot A \cdot C}}{2 \cdot A} \rightarrow \frac{-(-8) \pm \sqrt{(-8)^2 - 4 \cdot (1) \cdot (12)}}{2 \cdot (1)}$$

Plug the values into the formula

$$= \frac{-(-8) \pm \sqrt{(-8)^2 - 4 \cdot (1) \cdot (12)}}{2 \cdot (1)}$$

We strongly recommend the use of parenthesis to separate each component at first, it helps you keep track of the signs

$$= \frac{8 \pm \sqrt{64 - 48}}{2 \cdot (1)} = \frac{8 \pm \sqrt{16}}{2 \cdot (1)}$$

Start with the inside of the root

Double negative means it turns positive in the next step!

$$= \frac{8 \pm \sqrt{16}}{2 \cdot (1)}$$

$$= \frac{8 \pm 4}{2}$$

Solve the root and the multiplication in the denominator

$$= \frac{8 \pm 4}{2}$$

Split the plus or minus sign to get the two answers

$$x = \frac{8 + 4}{2} \qquad\qquad x = \frac{8 - 4}{2}$$

$$x = \frac{12}{2} \qquad\qquad x = \frac{4}{2}$$

$$x = 6 \qquad\qquad x = 2$$

$$x = 6 \text{ and } x = 2 \text{ »» } \textbf{\textit{Answers}}$$

Number in red is the operation performed in the current step
Number in blue is the operation performed in the previous step

Problem 63) $x^2 + 4x + 4 = 0$

$$x^2 + 4x + 4 = 0$$

$$1x^2 + 4x + 4 = 0$$

$$A = 1 \; ; B = 4 \; ; C = 4$$

Identify the values for A, B and C

$$\frac{-B \pm \sqrt{B^2 - 4 \cdot A \cdot C}}{2 \cdot A} \to \frac{-(4) \pm \sqrt{(4)^2 - 4 \cdot (1) \cdot (4)}}{2 \cdot (1)}$$

Plug the values into the formula

$$= \frac{-(4) \pm \sqrt{(4)^2 - 4 \cdot (1) \cdot (4)}}{2 \cdot (1)}$$

We strongly recommend the use of parenthesis to separate each component at first, it helps you keep track of the signs

$$= \frac{-4 \pm \sqrt{16 - 16}}{2 \cdot (1)} = \frac{-4 \pm \sqrt{0}}{2 \cdot (1)}$$

Start with the inside of the root

$$= \frac{-4 \pm \sqrt{0}}{2 \cdot (1)}$$

$$= \frac{-4 \pm 0}{2}$$

Solve the root and the multiplication in the denominator

$$= \frac{-4 \pm 0}{2}$$

Technically we still have two answers but since it is the same for both cases, we write it once. Knowing this will come handy in more advanced math courses

Split the plus or minus sign to get the two answers

$$x = \frac{-4 + 0}{2} \qquad x = \frac{-4 - 0}{2}$$

$$x = \frac{-4}{2} \qquad x = \frac{-4}{2}$$

$$x = -2 \qquad x = -2$$

$$x = -2 \; \text{»» } \textbf{\textit{Answer}}$$

Number in red is the operation performed in the current step
Number in blue is the operation performed in the previous step

<u>Quadratic equations: quadratic formula: practice problems</u>

1) $x^2 + 7x - 8 = 0$ 2) $x^2 - 15x + 50 = 0$

3) $x^2 + 2x = 8x + 7$ 4) $11x + 4 = -x^2 - 6$

5) $x^2 - 11x + 30 = 0$ 6) $x^2 - 2x - 195 = 0$

More quadratics!!

Well, let's have some more practice with quadratic equations as they tend to be problematic sometimes. We'll be using both factorization and quadratic formula methods to solve for the following examples:

Problem 64) $3x^2 + 10x + 8 = 0$

$$3x^2 + 10x + 8 = 0$$

$$3x^2 + 10x + 8 = 0$$

$$A = 3 \; ; B = 10 \; ; C = 8$$

Identify the values for A, B and C

- A is the number multiplying the x^2. If you don't see any, assume a 1
- B is the number multiplying the normal x. If you don't see any, assume a 1
- C is the number without x

$$A \cdot C = 3 \cdot 8 = 24$$

Multiply A and C together.

Now we must find two numbers that meet:

- When added each other are equal to B (10 in this problem)

$$6 + 4 = 10$$

- When multiplied each other are equal to A·C (24 in this problem)

$$(6) \cdot (4) = 24$$

So our magic numbers are <u>**positive 6**</u> and <u>**positive 4**</u>

$$3x^2 + 10x + 8 = 0$$

Split this number into the two magic numbers

$$3x^2 + 6x + 4x + 8 = 0$$

$$\{3x^2 + 6x\}\{+4x + 8\} = 0$$

Separate the new equation in two groups

A reminder about the common factor process

$$3x\left(\frac{3x^2}{3x} + \frac{6x}{3x}\right) =$$
$$3x(x + 2)$$

A reminder about the common factor process

$$4\left(\frac{4x}{4} + \frac{8}{4}\right) =$$
$$4(x + 2)$$

$$\{3x^2 + 6x\}\{+4x + 8\} = 0$$
③x ④

Take a common factor of 3x from the green (light) group and a 4 from the purple (dark) group

$$3x(x + 2) + 4(x + 2) = 0$$

$$3x(x + 2) + 4(x + 2) = 0$$

Equal values inside of the parenthesis, this is a good sign!!

Number in red is the operation performed in the current step

Number in blue is the operation performed in the previous step

Let's evaluate what we have:

$$3x(x + 2) + 4(x + 2) = 0$$

- Equal value inside of the parenthesis, in green = (x + 2)

- Leftovers outside of the parenthesis, in purple = (3x + 4)

- New equation resulting from multiplying both of the above: $(x + 2)(3x + 4) = 0$

We did it! The equation is now on factor form

$$(x + 2)(3x + 4) = 0$$

Apply the zero product property

$x + 2 = 0$ | $3x + 4 = 0$

$x + 2 = 0$ | $3x + 4 = 0$

Moves as negative		Moves as negative

$x = 0 - 2$ | $3x = 0 - 4$

$x = -2$ | | Moves as division |

$$x = \frac{-4}{3}$$

$$x = -2 \ \text{ and } \ x = -\frac{4}{3} \ \text{»» } Answers$$

Problem 65) $3x^2 + 10x + 8 = 0$

$$3x^2 + 10x + 8 = 0$$

$$3x^2 + 10x + 8 = 0$$

$$A = 3 \ ; B = 10; C = 8$$

Identify the values for A, B and C

$$\frac{-B \pm \sqrt{B^2 - 4 \cdot A \cdot C}}{2 \cdot A} \rightarrow \frac{-(10) \pm \sqrt{(10)^2 - 4 \cdot (3) \cdot (8)}}{2 \cdot (3)}$$

Plug the values into the formula

Number in red is the operation performed in the current step
Number in blue is the operation performed in the previous step

$$= \frac{-(10) \pm \sqrt{(10)^2 - 4 \cdot (3) \cdot (8)}}{2 \cdot (3)}$$

Start with the inside of the root

$$= \frac{-10 \pm \sqrt{100 - 96}}{2 \cdot (3)} = \frac{-10 \pm \sqrt{4}}{2 \cdot (3)}$$

$$= \frac{-10 \pm \sqrt{4}}{2 \cdot (3)}$$

$$= \frac{-10 \pm 2}{6}$$

Solve the root and the multiplication in the denominator

$$= \frac{-10 \pm 2}{6}$$

Split the plus or minus sign to get the two answers

$$x = \frac{-10 + 2}{6} \qquad\qquad x = \frac{-10 - 2}{6}$$

$$x = \frac{-8}{6} \qquad\qquad x = \frac{-12}{6}$$

$$x = -\frac{4}{3} \qquad\qquad x = -2$$

We always try to simplify our fractions as much as possible:

$$-\frac{8 \div 2}{6 \div 2} = -\frac{4}{3}$$

$$x = -\frac{4}{3} \; \text{ and } \; x = -2 \; \text{»» Answer}$$

Number in red is the operation performed in the current step
Number in blue is the operation performed in the previous step

75

Problem 66) $5x^2 + 13x - 6 = 0$

$$5x^2 + 13x - 6 = 0$$

$$5x^2 + 13x - 6 = 0$$

$$A = 5 \,; B = 13 \,; C = -6$$

Identify the values for A, B and C

- A is the number multiplying the x^2. If you don't see any, assume a 1
- B is the number multiplying the normal x. If you don't see any, assume a 1
- C is the number without x

$$A \cdot C = 5 \cdot -6 = -30$$

Multiply A and C together.

Now we must find two numbers that meet:

- When added each other are equal to B (13 in this problem)
$$15 - 2 = 13$$
When multiplied each other are equal to A·C (20 in this problem)
$$(15) \cdot (-2) = -30$$
So our magic numbers are **negative 10** and **negative 2**

$$5x^2 + 13x - 6 = 0$$

Split this number into the two magic numbers

$$5x^2 + 15x - 2x - 6 = 0$$

$$\{5x^2 + 15x\} \{-2x - 6\} = 0$$

Separate the new equation in two groups

A reminder about the common factor process

$$5x\left(\frac{5x^2}{5x} + \frac{15x}{5x}\right) = 5x(x + 3)$$

A reminder about the common factor process

$$-2\left(\frac{-2x}{-2} - \frac{6}{-2}\right) = -2(x + 3)$$

$$\{5x^2 + 15x\} \{-2x - 6\} = 0$$

Take a common factor of 5x from the green group and a -2 from the purple group

$$5x(x + 3) - 2(x + 3) = 0$$

$$5x(x + 3) - 2(x + 3) = 0$$

Equal values inside of the parenthesis, this is a good sign!!

Number in red is the operation performed in the current step
Number in blue is the operation performed in the previous step

Let's evaluate what we have:

$$5x(x + 3) - 2(x + 3) = 0$$

- Equal value inside of the parenthesis, in green = (x +3)

- Leftovers outside of the parenthesis, in purple = (5x - 2)

- New equation resulting from multiplying both of the above: (x+3)(5x-2)=0

We did it! The equation is now on factor form

$$(x + 3)(5x - 2) = 0$$

Apply the zero product property

$x + 3 = 0$ $5x - 2 = 0$

$x + 3 = 0$ $5x - 2 = 0$

Moves as negative   Moves as positive

$x = 0 \ {-3}$ $5x = 0 \ {+2}$

$x = -3$ Moves as division

$$x = \frac{2}{5}$$

$$x = -3 \ \text{and} \ x = \frac{2}{5} \ \text{»» Answers}$$

Problem 67) $5x^2 + 13x - 6 = 0$

$$5x^2 + 13x - 6 = 0$$

$$5x^2 + 13x - 6 = 0$$

$$A = 5 \ ; B = 13 \ ; C = -6$$

Identify the values for A, B and C

$$\frac{-B \pm \sqrt{B^2 - 4 \cdot A \cdot C}}{2 \cdot A} \rightarrow \frac{-(13) \pm \sqrt{(13)^2 - 4 \cdot (5) \cdot (-6)}}{2 \cdot (5)}$$

Plug the values into the formula

Number in red is the operation performed in the current step

Number in blue is the operation performed in the previous step

$$= \frac{-(13) \pm \sqrt{(13)^2 - 4 \cdot (5) \cdot (-6)}}{2 \cdot (5)}$$

Be careful!! There is a double negative in This operation!!!!

> Start with the inside of the root

$$= \frac{-13 \pm \sqrt{169 + 120}}{2 \cdot (5)} = \frac{-13 \pm \sqrt{289}}{2 \cdot (5)}$$

$$= \frac{-13 \pm \sqrt{289}}{2 \cdot (5)}$$

$$= \frac{-13 \pm 17}{10}$$

> Solve the root and the multiplication in the denominator

$$= \frac{-13 \pm 17}{10}$$

> Split the plus or minus sign to get the two answers

We always try to simplify our fractions as much as possible:

$$\frac{4 \div 2}{10 \div 2} = \frac{2}{5}$$

$$x = \frac{-13 + 17}{10} \qquad\qquad x = \frac{-13 - 17}{10}$$

$$x = \frac{4}{10} \qquad\qquad x = \frac{-30}{10}$$

$$x = \frac{2}{5} \qquad\qquad x = -3$$

$$x = \frac{2}{5} \ and \ x = -3 \ \text{»» Answers}$$

Number in red is the operation performed in the current step

Number in blue is the operation performed in the previous step

Problem 68) $5x^2 = 12x - 4$

$$5x^2 = 12x - 4$$

Moves as negative

$$5x^2 = 12x - 4$$

Let's make it look like a generic quadratic equation first!!

Moves as positive

$$5x^2 - 12x + 4 = 0$$

$$5x^2 - 12x + 4 = 0$$

Again, we just re-grouped everything on the same side of the equal sign. We told you it was a good technique

$$5x^2 - 12x + 4 = 0$$

$$A = 5 \; ; B = -12 \; ; C = 4$$

Identify the values for A, B and C

Now we can start with factorization

$$A \cdot C = 5 \cdot 4 = 20$$

Multiply A and C together.

Now we must find two numbers that meet:

- When added each other are equal to B (-12 in this problem)
$$-10 - 2 = -12$$
- When multiplied each other are equal to A·C (20 in this problem)
$$(-10) \cdot (-2) = 20$$

So our magic numbers are **negative 10** and **negative 2**

$$5x^2 - 12x + 4 = 0$$

Split this number into the two magic numbers

$$5x^2 - 10x - 2x + 4 = 0$$

$$\{5x^2 - 10x\}\{-2x + 4\} = 0$$

Separate the new equation in two groups

A reminder about the common factor process

$$5x\left(\frac{5x^2}{5x} - \frac{10x}{5x}\right) = 5x(x - 2)$$

A reminder about the common factor process

$$-2\left(\frac{-2x}{-2} + \frac{4}{-2}\right) = -2(x - 2)$$

$$\{5x^2 - 10x\}\{-2x + 4\} = 0$$

Take a common factor of 5x from the green (light) group and a -2 from the purple (dark) group

Number in red is the operation performed in the current step
Number in blue is the operation performed in the previous step

$$5x(x - 2) - 2(x - 2) = 0$$

$$5x(x - 2) - 2(x - 2) = 0$$

Equal values inside of the parenthesis, this is a good sign!!

Let's evaluate what we have:

$$5x(x - 2) - 2(x - 2) = 0$$

- Equal value inside of the parenthesis, in green = (x - 2)
- Leftovers outside of the parenthesis, in purple = (5x - 2)
- New equation resulting from multiplying both of the above: $(x - 2)(5x - 2) = 0$

We did it! The equation is now on factor form

$$(x - 2)(5x - 2) = 0$$

Apply the zero product property

$x - 2 = 0$	$5x - 2 = 0$
$x - 2 = 0$	$5x - 2 = 0$
Moves as positive	Moves as positive
$x = +2$	$5x = 0 + 2$
$x = 2$	Moves as division
	$x = \dfrac{2}{5}$

$$x = 2 \ \text{ and } \ x = \frac{2}{5} \ \text{»» Answers}$$

Number in red is the operation performed in the current step
Number in blue is the operation performed in the previous step

80

Problem 69) $5x^2 = 12x - 4$

$$5x^2 = 12x - 4$$

Moves as negative

$$5x^2 = 12x - 4$$

Moves as positive

Same as factorization, we have to make it look like a generic quadratic

Everything was moved to one side of the equal sign, as usual.

$$5x^2 - 12x + 4 = 0$$

$$5x^2 - 12x + 4 = 0$$

$$5x^2 - 12x + 4 = 0$$

As we made it into a generic quadratic, we can proceed with the quadratic formula

$$A = 5 \; ; B = -12 ; C = 4$$

Identify the values for A, B and C

$$\frac{-B \pm \sqrt{B^2 - 4 \cdot A \cdot C}}{2 \cdot A} \rightarrow \frac{-(-12) \pm \sqrt{(-12)^2 - 4 \cdot (5) \cdot (4)}}{2 \cdot (5)}$$

Plug the values into the formula

$$= \frac{-(-12) \pm \sqrt{(-12)^2 - 4 \cdot (5) \cdot (4)}}{2 \cdot (5)}$$

Double negatives!! Double negatives!!!

Start with the inside of the root

$$= \frac{12 \pm \sqrt{144 - 80}}{2 \cdot (5)} = \frac{12 \pm \sqrt{64}}{2 \cdot (5)}$$

$$= \frac{12 \pm \sqrt{64}}{2 \cdot (5)}$$

$$= \frac{12 \pm 8}{10}$$

Solve the root and the multiplication in the denominator

Number in red is the operation performed in the current step
Number in blue is the operation performed in the previous step

81

$$= \frac{12 \pm 8}{10}$$

Split the plus or minus sign to get the two answers

A reminder about the common factor process

$$-2\left(\frac{-2x}{-2} + \frac{4}{-2}\right)$$
$$= -2(x - 2)$$

$$x = \frac{12 + 8}{10} \qquad x = \frac{12 - 8}{10}$$

$$x = \frac{20}{10} \qquad x = \frac{4}{10}$$

$$x = 2 \qquad x = \frac{2}{5}$$

$$x = 2 \ and \ x = \frac{2}{5} \ \text{\textguillemetright\textguillemetright} \ Answers$$

Problem 70) $2x^2 + 4x + 5 = 0$

If you spend as much time as we did trying to solve this problem by factorization you would probably realize it is waaaaaay too complicated to find the two magic numbers. We could probably try the quadratic equation to see if we can find and answer but this is a good example to show how can we prove whether a quadratic equation has a solution or not. For this we will use what is called "the discriminant":

The discriminant is just the inside of the root in the quadratic equation:
$$\frac{-B \pm \sqrt{B^2 - 4 \cdot A \cdot C}}{2 \cdot A} \rightarrow Discriminant = B^2 - 4 \cdot A \cdot C$$

We will just place the respective numbers in the discriminant equation, if our answer is a negative number (A.K.A smaller than zero), it means the equation has no solution.

$$2x^2 + 4x + 5 = 0$$

$$A = 2 \ ; B = 4 \ ; C = 5$$

Identify the values for A, B and C

$$D = (4)^2 - 4 \cdot (2) \cdot (5)$$

$$D = 16 - 40$$

$$D = -24$$

$$-24 < 0$$

The discriminant is smaller than zero, therefore this quadratic equation has no real solution.

No Answer

P.S. There is such a thing as an *imaginary* solution, but we will review it in future books

Number in red is the operation performed in the current step

Number in blue is the operation performed in the previous step

Note about the discriminant: The discriminant is an operation we use in quadratic equations to find whether it has two solutions, one solution or no solution at all. To find the discriminant of a quadratic equation we need to take a look at the quadratic formula:

$$\frac{-B \pm \sqrt{B^2 - 4 \cdot A \cdot C}}{2 \cdot A}$$

The discriminant is the operation performed inside of the square root:

$$B^2 - 4 \cdot A \cdot C = \text{Discriminant}$$

The value obtained will determine the number of solutions the equation may have. Let's take a look at the following table to see all the possibilities:

Value of the discriminant	Number of solutions of the equation
Any positive number	Two real solutions
Zero	One real solution
Any negative number	Zero real solution (possible imaginary solution)

More quadratics!!: practice problems

1) $5x^2 + 3x - 2 = 0$ 2) $6x^2 - 17x - 3 = 0$

3) $9x^2 - 12 = 3x^2 - 25x - 3$ 4) $28x^2 + 6 = 30 - 11x$

5) $2x^2 - 19x - 10 = 0$ 6) $6x^2 - 5x + 11 = 17$

Number in red is the operation performed in the current step
Number in blue is the operation performed in the previous step

83

Higher degree polynomials

As you probably guessed, yes, there are equations even harder than quadratic equations. Many times our highest exponent will not be a nice number two, it will be larger than that. At this point, the quadratic methods no longer apply.

If we take a close look at the factorization method for the quadratic equations, all we are doing is reducing its exponents until we get two separate equations both with an x having an exponent of 1. A similar situation will happen with these higher degree polynomials. We will try to reduce them until we have different equations with lower exponents, easier to solve. It is important to remember the common factor rule and the zero product property.

Also, it will be very important to know a couple more facts before we jump into these harder problems:

- The methods used to reduce the fractions are called "polynomial division" and "synthetic division". We will learn both in this book. The methods will be explained in each example.

- We need to know that the zero of a function is any number, when plugged in for x, gives zero as the final answer. For example

For the equation:

$$3x^3 - 4x^2 + 5x - 4$$

$$x = 1 \; is \; a \; zero$$

As we plug it in:

$$3(1)^3 - 4(1)^2 + 5(1) - 4 = 3 - 4 + 5 - 4 = 0$$

- Many times it will not be easy to find the zero of a specific equation, especially as we have to use trial and error. There is a trick we can use to narrow our search:

For the equation:

$$3x^3 - 4x^2 + 5x - 4$$

Multiply the first and last number (sign doesn't matter)

$$3 \cdot 4 = 12$$

Find all factors of the multiplication, both positive and negative:

$$Factors \; of \; 12: \pm 1; \; \pm 2; \; \pm 3; \; \pm 4; \; \pm 6; \; \pm 12$$

One of those factors will most likely be a zero, as in this case, positive 1 was a zero. It is still a pretty wide range to look for, but at least it is smaller than infinity.

Number in red is the operation performed in the current step
Number in blue is the operation performed in the previous step

Problem 71) $2x^3 - 7x^2 - 5x + 4 = 0$

$$2x^3 - 7x^2 - 5x + 4 = 0$$

Find the initial zero

$$2x^3 - 7x^2 - 5x + 4 = 0$$

Multiply the first and last numbers

$$2 \cdot 4 = 8$$

Find all factors of the answer

$$Factors \ of \ 8 = \pm1; \ \pm2; \ \pm4; \ \pm8$$

Try each factor on the equation until it gives an answer of 0

$For \ this \ equation, x = -1 \ is \ a \ zero$

$$2(-1)^3 - 7(-1)^2 - 5(-1) + 4$$

$$2(-1) - 7(1) - 5(-1) + 4$$

$$-2 - 7 + 5 + 4 = 0$$

Yes, x $= -1$ is a zero!

We kind of cheated, since we created the problem we already knew x=-1 was a zero, but get some practice on this, it is the most tedious part of the process

Now that we have our initial zero, we must set up the:

Polynomial Division:

1: Here we place the value we found as or initial zero. Very important to change its sign.
2: This is where our answer will be showing

$$02$$

$$x + 1 \overline{\smash{\big)}\ 2x^3 - 7x^2 - 5x + 4}$$

$$01 \qquad \qquad 03$$

3: In this position we place the equation we are trying to reduce

Number in red is the operation performed in the current step
Number in blue is the operation performed in the previous step

2nd step:

1: Divide the element in the blue square by the element in the thick red one:
$$\frac{2x^3}{x} = 2x^2$$
2: Place the answer of step 1 here

$$x + 1 \quad \boxed{2x^3 - 7x^2} - 5x + 4$$
$$2x^3 + 2x^2$$
$$-9x^2 - 5x$$
$$2x^2$$

3: Multiply the answer of step 1 by the x+1 at the "zero" location
4: Place the answer of step 3 in this position
5: Subtract elements in the dashed rectangle
6: Place the answer of step 5 in this position
7: Move the number In red to the position with the number in blue

$$(x + 1) \cdot 2x^2 = x \cdot 2x^2 + 1 \cdot 2x^2 = \mathbf{2x^3 + 2x^2}$$

$$(2x^3 - 7x^2) - (2x^3 + 2x^2) = 2x^3 - 7x^2 - 2x^3 - 2x^2$$
$$= 2x^3 - 2x^3 - 7x^2 - 2x^2 = \mathbf{9x^2}$$

3rd step:

1: Divide the element in the blue square by the element in the thick red one:
$$\frac{-9x^2}{x} = -9x$$
2: Place the answer of step 1 here

$$x + 1 \quad \begin{array}{l} 2x^2 - 9x + 4 \\ \hline 2x^3 - 7x^2 - 5x + 4 \\ 2x^3 + 2x^2 \\ \quad -9x^2 - 5x \\ \quad -9x^2 - 9x \\ \qquad\quad 4x + 4 \end{array}$$

3: Multiply the answer of step 1 by the x+1 at the "zero" location
4: Place the answer of step 3 in this position
5: Subtract elements in the dashed rectangle
6: Place the answer of step 5 in this position
7: Move the number in red to the position with the number in blue

$$(x + 1) \cdot -9x = x \cdot -9x + 1 \cdot -9x = \mathbf{-9x^2 - 9x}$$

$$(-9x^2 - 5x) - (-9x^2 - 9x) = -9x^2 - 5x + 9x^2 + 9x$$
$$= -9x^2 + 9x^2 - 5x + 9x = \mathbf{4x}$$

Number in red is the operation performed in the current step
Number in blue is the operation performed in the previous step

4th step:

1: Divide the element in the blue square by the element in the red one:
$$\frac{4x}{x} = 4$$
2: Place the answer of step 1 here

3: Multiply the answer of step 1 by the x+1 at the "zero" location
4: Place the answer of step 3 in this position
5: Subtract elements in the yellow rectangle
6: Place the answer of step 5 in this position
7: Our final subtraction is equal to 0 and there are no more numbers to bring down. The division is over

$$
\begin{array}{r}
2x^2 - 9x + 4 \\
x+1 \enclose{longdiv}{2x^3 - 7x^2 - 5x + 4} \\
2x^3 + 2x^2 \\
-9x^2 - 5x \\
-9x^2 - 9x \\
4x + 4 \\
4x + 4 \\
0 + 0
\end{array}
$$

$$(x + 1) \cdot 4 = x \cdot 4 + 1 \cdot 4 = \mathbf{4x + 4}$$

$$(4x + 4) - (4x + 4) = 4x + 4 - 4x - 4$$
$$= 4x - 4x + 4 - 4 = \mathbf{0 + 0}$$

Remember the first step of the division. Our answer is showing at the top of the division. Our reduced equation is the multiplication of our zero and the answer of the division:

$$2x^3 - 7x^2 - 5x + 4 \rightarrow Reduced\ to \rightarrow (x + 1)(2x^2 - 9x + 4)$$

Number in red is the operation performed in the current step
Number in blue is the operation performed in the previous step

In order to complete the solution, we have to set the equation equal to 0 as it originally was written

$$(x+1)\left(2x^2 - 9x + 4\right) = 0$$

Apply the zero product property

$x + 1 = 0$ | $2x^2 - 9x + 4 = 0$

$x + 1 = 0$

$2x^2 - 9x + 4 = 0$

Moves as negative

$A = 2 \; ; B = -9 \; ; C = 4$

Identify the values of A, B and C

$x = 0^{-1}$

$$x = \frac{-(-9) \pm \sqrt{(-9)^2 - 4 \cdot (2) \cdot (4)}}{2 \cdot (2)}$$

$x = -1$

$$x = \frac{9 \pm \sqrt{81 - 32}}{4}$$

$$x = \frac{9 \pm \sqrt{49}}{4}$$

$$x = \frac{9 + 7}{4} \quad ; \quad x = \frac{9 - 7}{4}$$

$$x = \frac{16}{4} \quad ; \quad x = \frac{2}{4}$$

$$x = 4 \quad ; \quad x = \frac{1}{2}$$

$$x = -1 \; ; x = 4 \; and \; x = \frac{1}{2} \; \text{»» Answers}$$

After such a long process you might have forgotten how to simplify so we will remember it for you

$$\frac{2 \div 2}{4 \div 2} = \frac{1}{2}$$

Number in red is the operation performed in the current step

Number in blue is the operation performed in the previous step

Problem 71-b) $2x^3 - 7x^2 - 5x + 4 = 0$

$$2x^3 - 7x^2 - 5x + 4 = 0$$

<u>Synthetic division:</u>

Keep in mind from previous steps we already know the following:

$$x = -1 \ is \ a \ zero$$

The first step is to set up the division:

1: This is the position where we place our zero. For the synthetic division we don't change its sign
2: This is where our answer will be showing

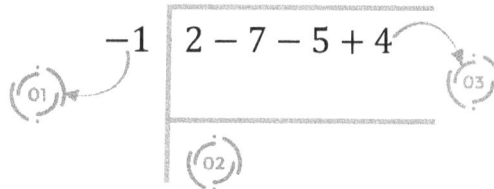

3: Here we will place the coefficients of the equation we are trying to reduce:
For the equation:

$2x^3 - 7x^2 - 5x + 4$

Its coefficients are:

$2 - 7 - 5 + 4$

2nd step:

1: Bring down the number colored in red to the position of the number in blue
2: Multiply the number in the thick red box with the number in the blue box
3: Place the answer of step two in this position

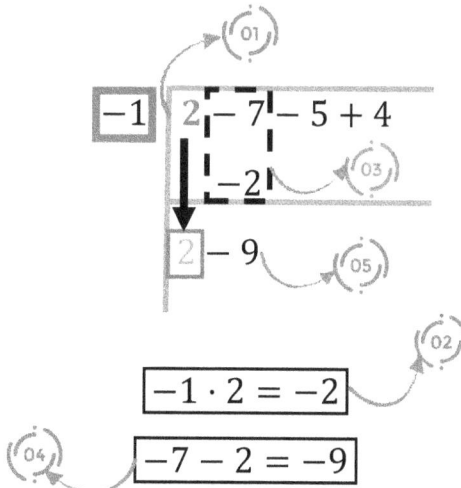

$$-1 \cdot 2 = -2$$

$$-7 - 2 = -9$$

4: Add the numbers in the dashed rectangle. Remember in math, the word "add" means addition or subtraction depending on the signs
5: Write the answer of step 4 in this position

Number in red is the operation performed in the current step
Number in blue is the operation performed in the previous step

3rd step:

1: Multiply the number in the thick red box with the number in the blue box
2: Place the answer of step one in this position

$$\boxed{-1} \quad 2 - 7 - 5 + 4$$
$$-2 + 9 \quad (02)$$
$$2 - 9 + 4 \quad (04)$$

3: Add the numbers in the dashed rectangle. Remember in math, the word "add" means addition or subtraction depending on the signs
4: Write the answer of step 3 in this position

(03) $\boxed{-1 \cdot -9 = 9}$ (01)

$\boxed{-5 + 9 = 4}$

4th step:

1: Multiply the number in the thick red box with the number in the blue box
2: Place the answer of step one in this position

$$\boxed{-1} \quad 2 - 7 - 5 + 4 \quad (02)$$
$$-2 + 9 - 4$$
$$2 - 9 + 4 + 0 \quad (04)$$

3: Add the numbers in the dashed rectangle. Remember in math, the word "add" means addition or subtraction depending on the signs
4: Write the answer of step 3 in this position

(01) $\boxed{-1 \cdot 4 = -4}$

(03) $\boxed{4 - 4 = 0}$

Number in red is the operation performed in the current step
Number in blue is the operation performed in the previous step

Remember from the first step where our answer will be showing. One more thing, the synthetic divisions gives us the answer in terms of coefficients:

$$Coefficients\ of\ the\ answer = 2 - 9 + 4$$

$$Our\ reduced\ answer = 2x^2 - 9x + 4$$

Our initial equation had a highest exponent of x^3 therefore our reduced equation MUST start with x^2

Hint: If the last number of the synthetic division (the remainder) is NOT zero, then we did something wrong

Same as the polynomial division,

the final reduced solution from which we can solve

Is the multiplication of the synthetic answer and the initial zero:

$$(x + 1)(2x^2 - 9x + 4)$$

Number in red is the operation performed in the current step
Number in blue is the operation performed in the previous step

91

In order to complete the solution, we have to set the equation equal to 0 as it originally was written

$$(x + 1)\left(2x^2 - 9x + 4\right) = 0$$

Apply the zero product property

$x + 1 = 0$ | $2x^2 - 9x + 4 = 0$

$x + 1 = 0$ | $2x^2 - 9x + 4 = 0$

Moves as negative | $A = 2 \; ; B = -9 \; ; C = 4$

Identify the values of A, B and C

$x = 0 \,^{-1}$ | $$x = \frac{-(-9) \pm \sqrt{(-9)^2 - 4 \cdot (2) \cdot (4)}}{2 \cdot (2)}$$

$x = -1$ | $$x = \frac{9 \pm \sqrt{81 - 32}}{4}$$

Yes, you are right! This page was just a big copy and paste from the solution of the polynomial division part, but it is literally the same thing!!

$$x = \frac{9 \pm \sqrt{49}}{4}$$

$$x = \frac{9 + 7}{4} \quad ; \quad x = \frac{9 - 7}{4}$$

$$x = \frac{16}{4} \quad ; \quad x = \frac{2}{4}$$

$$x = 4 \quad ; \quad x = \frac{1}{2}$$

$$x = -1 \; ; x = 4 \; and \; x = \frac{1}{2} \;\;»» \textbf{\textit{Answers}}$$

Number in red is the operation performed in the current step
Number in blue is the operation performed in the previous step

92

Problem 72) $3x^3 - 5x^2 + 4x - 12 = 0$

$$3x^3 - 5x^2 + 4x - 12 = 0$$

Find the initial zero

$$3x^3 - 5x^2 + 4x - 12 = 0$$

Multiply the first and last numbers

$$3 \cdot 12 = 36$$

Find all factors of the answer

Yes, this is still a big list of numbers but it is definitely smaller than infinity

$$Factors\ of\ 36 = \pm1;\ \pm2;\ \pm3;\ \pm4;\ \pm6;\ \pm9;\ \pm12;\ \pm18;\ \pm36$$

Try each factor on the equation until it gives an answer of 0

For this equation, $x = 2$ is a zero

$$3(2)^3 - 5(2)^2 + 4(2) - 12$$
$$3(8) - 5(4) + 4(2) - 12$$
$$24 - 20 + 8 - 12 = 0$$

Yes, $x = 2$ is a zero!

Again, we already knew what the zero was, I mean, we invented the problem! But you should give it a try on your own

Now that we have our initial zero we must set up the: <u>Polynomial Division:</u>

Number in red is the operation performed in the current step
Number in blue is the operation performed in the previous step

1st step:

1: Here we place the value we found as or initial zero. Very important to change its sign.
2: This is where our answer will be showing

(02)

$$x - 2 \mid \overline{3x^3 - 5x^2 + 4x - 12}$$

(01)   (03)

3: In this position we place the equation we are trying to reduce

2nd step:

1: Divide the element in the blue square by the element in the thick red one:
$$\frac{3x^3}{x} = 3x^2$$
2: Place the answer of step 1 here

$$
\boxed{x} - 2 \quad \boxed{3x^3 - 5x^2} + 4x - 12
$$

$$3x^2$$ (02)

$$3x^3 - 6x^2$$

$$x^2 + 4x$$

(04)   (06)

3: Multiply the answer of step 1 by the x-2 at the "zero" location
4: Place the answer of step 3 in this position
5: Subtract elements in the dashed rectangle
6: Place the answer of step 5 in this position
7: Move the number in red to the position with the number in blue

(03)

$$\boxed{(x - 2) \cdot 3x^2 = x \cdot 3x^2 - 2 \cdot 3x^2 = \mathbf{3x^3 - 6x^2}}$$

$$\boxed{\begin{aligned}(3x^3 - 5x^2) - (3x^3 - 6x^2) &= 3x^3 - 5x^2 - 3x^3 + 6x^2 \\ &= 3x^3 - 3x^3 - 5x^2 + 6x^2 = \mathbf{x^2}\end{aligned}}$$

(05)

3rd step:

$3x^2 + x$ ⟨02⟩

$x - 2$ ⟌ $3x^3 - 5x^2 + 4x - 12$
$3x^3 - 6x^2$

$\boxed{x^2} - 4x$
$x^2 - 2x$

$6x - 12$

1: Divide the element in the blue square by the element in the thick red one:
$$\frac{x^2}{x} = x$$
2: Place the answer of step 1 here

3: Multiply the answer of step 1 by the x-2 at the "zero" location
4: Place the answer of step 3 in this position
5: Subtract elements in the dashed rectangle
6: Place the answer of step 5 in this position
7: Move the number in red to the position with the number in blue

$$(x - 2) \cdot x = x \cdot x - 2 \cdot x = \boldsymbol{x^2 - 2x}$$

$$(x^2 + 4x) - (x^2 - 2x) = x^2 + 4x - x^2 + 2x$$
$$= x^2 - x^2 + 4x + 2x = \boldsymbol{6x}$$

4th step:

$3x^2 + x + 6$ ⟨02⟩

$x - 2$ ⟌ $3x^3 - 5x^2 + 4x - 12$
$3x^3 - 6x^2$

$x^2 + 4x$

$x^2 - 2x$

$\boxed{6x} - 12$
$6x - 12$
$0 + 0$

1: Divide the element in the blue square by the element in the thick red one:
$$\frac{6x}{x} = 6$$
2: Place the answer of step 1 here

3: Multiply the answer of step 1 by the x-2 at the "zero" location
4: Place the answer of step 3 in this position
5: Subtract elements in the dashed rectangle
6: Place the answer of step 5 in this position
7: Our final subtraction is equal to 0 and there are no more numbers to bring down. The division is over.

$$(x - 2) \cdot 6 = x \cdot 6 - 2 \cdot 6 = \boldsymbol{6x - 12}$$

$$(6x - 12) - (6x - 12) = 6x - 12 - 6x + 12$$
$$= 6x - 6x - 12 + 12 = \boldsymbol{0 + 0}$$

Number in red is the operation performed in the current step
Number in blue is the operation performed in the previous step

Remember the first step of the division. Our answer is showing at the top of the division

Our reduced equation is the multiplication of our zero and the answer of the division:

$$3x^3 - 5x^2 + 4x - 12 \rightarrow Reduced\ to \rightarrow (x-2)(3x^2 + x + 6)$$

In order to complete the solution, we have to set the equation equal to 0 as it originally was written

$$(x-2)(3x^2 + x + 6) = 0$$

Apply the zero product property

$$x - 2 = 0 \qquad\qquad 3x^2 + x + 6 = 0$$

$$x - 2 = 0 \qquad\qquad 3x^2 + 1x + 6 = 0$$

Moves as positive

$$A = 3\ ; B = 1\ ; C = 6$$

Identify the values of A, B and C

$$x = 0 + 2 \qquad\qquad x = \frac{-(1) \pm \sqrt{(1)^2 - 4 \cdot 3 \cdot 6}}{2 \cdot (3)}$$

$$x = 2 \qquad\qquad x = \frac{-1 \pm \sqrt{1 - 72}}{6}$$

$$x = \frac{-1 \pm \sqrt{-71}}{6}$$

We are facing the square root of a negative number, which, a mentioned before, DO NOT exists, therefore this side does NOT provide any answer to the problem

$$x = 2 \quad \text{»» } \textbf{\textit{Answer}}$$

Number in red is the operation performed in the current step
Number in blue is the operation performed in the previous step

96

Problem 72-b) $3x^3 - 5x^2 + 4x - 12 = 0$

$$3x^3 - 5x^2 + 4x - 12 = 0$$

<u>Synthetic division:</u>

Keep in mind from previous steps we already know the following:

$$x = 2 \ is \ a \ zero$$

The **1st step** is to set up the division:

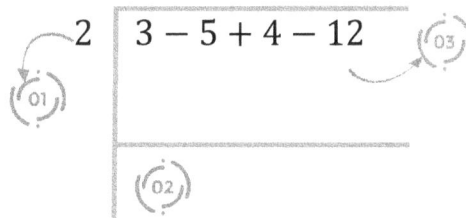

1: This is the position where we place our zero. For the synthetic division we don't change its sign.
2: This is where our answer will be showing

$$2 \ | \ 3 - 5 + 4 - 12$$

3: Here we will place the coefficients of the equation we are trying to reduce:

For the equation:

$$3x^3 - 5x^2 + 4x - 12$$

Its coefficients are:

$$3 - 5 + 4 - 12$$

2nd step:

$$\boxed{2} \ | \ 3 - 5 + 4 - 12$$
$$6$$
$$3 + 1$$

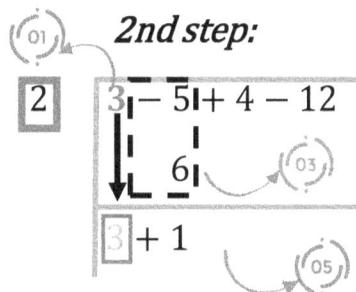

1: Bring down the number colored in red to the position of the number in blue
2: Multiply the number in the thick red box with the number in the blue box
3: Place the answer of step two in this position

4: Add the numbers in the dashed rectangle. Remember in math, the word "add" means addition or subtraction depending on the signs
5: Write the answer of step 4 in this position.

$$\boxed{2 \cdot 3 = 6}$$

$$\boxed{-5 + 6 = 1}$$

Number in red is the operation performed in the current step
Number in blue is the operation performed in the previous step

97

3rd step:

2 | $3 - 5 + 4 - 12$

$6 + 2$

$3 + 1 + 6$

$2 \cdot 1 = 2$

$4 + 2 = 6$

1: Multiply the number in the thick red box with the number in the blue box
2: Place the answer of step one in this position

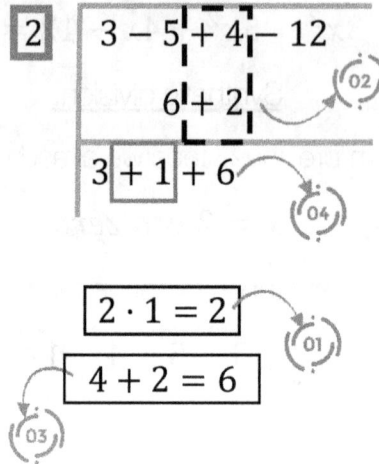

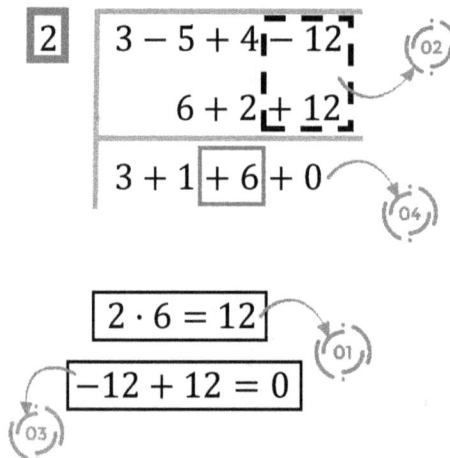

3: Add the numbers in the dashed rectangle. Remember in math, the word "add" means addition or subtraction depending on the signs
4: Write the answer of step 3 in this position

4th step:

2 | $3 - 5 + 4 - 12$

$6 + 2 + 12$

$3 + 1 + 6 + 0$

$2 \cdot 6 = 12$

$-12 + 12 = 0$

1: Multiply the number in the thick red box with the number in the blue box
2: Place the answer of step one in this position

3: Add the numbers in the dashed rectangle. Remember in math, the word "add" means addition or subtraction depending on the signs
4: Write the answer of step 3 in this position

Remember from the first step where our answer will be showing. One more thing, the synthetic divisions gives us the answer in terms of coefficients:

$$Coefficients\ of\ the\ answer = 3 + 1 + 6$$
$$Our\ reduced\ answer = 3x^2 + 1x + 6$$

Hint: If the last number of the synthetic division is NOT zero, then we did something wrong

Same as the polynomial division,

the final reduced solution from which we can solve

Our initial equation had a highest exponent of x^3 therefore our reduced equation MUST start with x^2

Is the multiplication of the synthetic answer and the initial zero:

$$(x + 1)(2x^2 - 9x + 4)$$

Number in red is the operation performed in the current step
Number in blue is the operation performed in the previous step

In order to complete the solution, we have to set the equation equal to 0 as it originally was written

$$(x - 2)(3x^2 + x + 6) = 0$$

Apply the zero product property

$x - 2 = 0$ $3x^2 + x + 6 = 0$

$x - 2 = 0$ $3x^2 + 1x + 6 = 0$

Moves as positive

$A = 3; B = 1 ; C = 6$

Identify the values of A, B and C

$$x = 0 + 2$$

$$x = \frac{-(1) \pm \sqrt{(1)^2 - 4 \cdot 3 \cdot 6}}{2 \cdot (3)}$$

$$x = 2$$

$$x = \frac{-1 \pm \sqrt{1 - 72}}{6}$$

$$x = \frac{-1 \pm \sqrt{-71}}{6}$$

We are facing the square root of a negative number, which, a mentioned before, DO NOT exists, therefore this side does NOT provide any answer to the problem !

$$x = 2 \quad »» \textbf{\textit{Answer}}$$

Number in red is the operation performed in the current step
Number in blue is the operation performed in the previous step

Problem 73) $x^3 - 10x + 3 = 0$

$$x^3 - 10x + 3 = 0$$

> Find the initial zero

$$1x^3 - 10x + 3 = 0$$

> Multiply the first and last numbers

$$1 \cdot 3 = 3$$

> Find all factors of the answer

$$Factors\ of\ 3 = \pm 1;\ \pm 3$$

> Try each factor on the equation until it gives an answer of 0

$$For\ this\ equation, x = 3\ is\ a\ zero$$

$$(3)^3 - 10(3) + 3$$

$$(27) - 10(3) + 3$$

$$27 - 30 + 3 = 0$$

Again, we already knew what the zero was, I mean, we invented the problem! But you should give it a try on your own

$$Yes, x = 3\ is\ a\ zero!$$

Now that we have our initial zero,
we must set up the: **Polynomial Division:**

Number in red is the operation performed in the current step
Number in blue is the operation performed in the previous step

100

1st step:

1: Here we place the value we found as or initial zero. Very important to change its sign.
2: This is where our answer will be showing

3: In this position we place the equation we are trying to reduce
4: All powers of x must be included for the division to work. We were missing x^2 in the original so we include it multiplying times 0

$$x - 3 \; \big| \; \boxed{x^3} + 0x^2 - 10x + 3$$

2nd step:

1: Divide the element in the blue square by the element in the thick red one:
$$\frac{x^3}{x} = x^2$$
2: Place the answer of step 1 here

3: Multiply the answer of step 1 by the x-3 at the "zero" location
4: Place the answer of step 3 in this position
5: Subtract elements in the dashed rectangle
6: Place the answer of step 5 in this position
7: Move the number in red to the position with the number in blue

$$\boxed{x} - 3 \; \big| \; \begin{array}{l} x^2 \\ \boxed{x^3} + 0x^2 - 10x + 3 \\ x^3 - 3x^2 \\ \hline 3x^2 - 10x \end{array}$$

$$(x - 3) \cdot x^2 = x \cdot x^2 - 3 \cdot x^2 = \mathbf{x^3 - 3x^2}$$

$$(x^3 + 0x^2) - (x^3 - 3x^2) = x^3 + 0x^2 - x^3 + 3x^2$$
$$= \cancel{x^3} - \cancel{x^3} + 0x^2 + 3x^2 = \mathbf{3x^2}$$

Number in red is the operation performed in the current step
Number in blue is the operation performed in the previous step

101

3rd step:

1: Divide the element in the blue square by the element in the thick red one:

$$\frac{3x^2}{x} = 3x$$

2: Place the answer of step 1 here

$\boxed{x} - 3$

$$x^2 + 3x$$

$$
\begin{array}{r}
x^3 + 0x^2 - 10x + 3 \\
x^3 - 3x^2 \\
\hline
\boxed{3x^2} - 10x \\
3x^2 - 9x \\
\hline
-x + 3
\end{array}
$$

3: Multiply the answer of step 1 by the x-3 at the "zero" location
4: Place the answer of step 3 in this position
5: Subtract elements in the dashed rectangle
6: Place the answer of step 5 in this position
7: Move the number in red to the position with the number in blue

$$\boxed{(x - 3) \cdot 3x = x \cdot 3x - 3 \cdot 3x = 3x^2 - 9x}$$

$$\boxed{\begin{aligned}(3x^2 - 10x) - (3x^2 - 9x) &= 3x^2 - 10x - 3x^2 + 9x \\ &= 3x^2 - 3x^2 - 10x + 9x = -x\end{aligned}}$$

4th step:

1: Divide the element in the blue square by the element in the red one:

$$\frac{-x}{x} = -1$$

2: Place the answer of step 1 here

$\boxed{x} - 3$

$$x^2 + 3x - 1$$

$$
\begin{array}{r}
x^3 + 0x^2 - 10x + 3 \\
x^3 - 3x^2 \\
3x^2 - 10x \\
3x^2 - 9x \\
\hline
\boxed{-x} + 3 \\
-x + 3 \\
\hline
0 + 0
\end{array}
$$

3: Multiply the answer of step 1 by the x-3 at the "zero" location
4: Place the answer of step 3 in this position
5: Subtract elements in the dashed rectangle
6: Place the answer of step 5 in this position
7: Our final subtraction is equal to 0 and there are no more numbers to bring down. The division is over

$$\boxed{(x - 3) \cdot -1 = x \cdot -1 - 3 \cdot -1 = -x + 3}$$

$$\boxed{\begin{aligned}(-x + 3) - (-x + 3) &= -x + 3 + x - 3 \\ &= -x + x + 3 - 3 = 0 + 0\end{aligned}}$$

Number in red is the operation performed in the current step
Number in blue is the operation performed in the previous step

Remember the first step of the division. Our answer is showing at the top of the division
Our reduced equation is the multiplication of our zero and the answer of the division:

$$x^3 - 10x + 3 \rightarrow Reduced\ to \rightarrow (x-3)(x^2 + 3x - 1)$$

In order to complete the solution, we have to set the equation equal to 0 as it originally was written

$$(x-3)(x^2 + 3x - 1) = 0$$

Apply the zero product property

$$x - 3 = 0 \qquad\qquad x^2 + 3x - 1 = 0$$

$$x - 3 = 0 \qquad\qquad 1x^2 + 3x - 1 = 0$$

Moves as positive

$$A = 1\ ; B = +3\ ; C = -1$$

Identify the values of A, B and C

$$x = 0 +3 \qquad\qquad x = \frac{-(3) \pm \sqrt{(3)^2 - 4 \cdot (1) \cdot (-1)}}{2 \cdot (1)}$$

$$x = 3 \qquad\qquad x = \frac{-3 \pm \sqrt{9 + 4}}{2}$$

$$x = \frac{-3 \pm \sqrt{13}}{2}$$

$$x = \frac{-3 + \sqrt{13}}{2} \quad ; \quad x = \frac{-3 - \sqrt{13}}{2}$$

These last two answers have a large number of decimals. We can just leave them in this form, it is correct and it is also cleaner

$$x = 3\ ; x = \frac{-3 + \sqrt{13}}{2}\ and\ x = \frac{-3 - \sqrt{13}}{2} \quad »» \ Answers$$

Number in red is the operation performed in the current step
Number in blue is the operation performed in the previous step

Problem 73-b) $x^3 - 10x + 3 = 0$

$$x^3 - 10x + 3 = 0$$

Synthetic division:

Keep in mind from previous steps we already know the following:

$$x = 3 \text{ is a zero}$$

The **1st step** is to set up the division:

1: This is the position where we place our zero. For the synthetic division we don't change its sign
2: This is where our answer will be showing

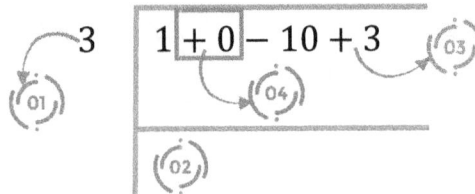

$$3 \quad | \quad 1 + 0 - 10 + 3$$

3: Here we will place the coefficients of the equation we are trying to reduce:
For the equation:
$1x^3 + 0x^2 - 10x + 3$
Its coefficients are:
$1 + 0 - 10 + 3$
4: Same as polynomial division, any missing power must be compensated with a zero

2nd step:

1: Bring down the number colored in red to the position of the number in blue
2: Multiply the number in the thick red box with the number in the blue box
3: Place the answer of step two in this position

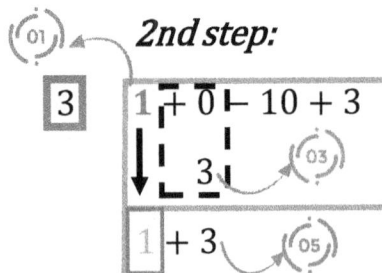

$$3 \quad | \quad 1 + 0 - 10 + 3$$
$$ 3$$
$$ 1 + 3$$

4: Add the numbers in the dashed rectangle. Remember in math, the word "add" means addition or subtraction depending on the signs
5: Write the answer of step 4 in this position

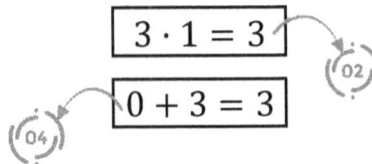

$$3 \cdot 1 = 3$$
$$0 + 3 = 3$$

Number in red is the operation performed in the current step
Number in blue is the operation performed in the previous step

104

3rd step:

1: Multiply the number in the thick red box with the number in the blue box
2: Place the answer of step one in this position

$$3 \mid 1 + 0 - 10 + 3$$
$$3 + 9$$
$$1 + 3 - 1$$

$$3 \cdot 3 = 9$$
$$-10 + 9 = -1$$

3: Add the numbers in the dashed rectangle. Remember in math, the word "add" means addition or subtraction depending on the signs
4: Write the answer of step 3 in this position

4th step:

1: Multiply the number in the thick red box with the number in the blue box
2: Place the answer of step one in this position

$$3 \mid 1 + 0 - 10 + 3$$
$$3 + 9 - 3$$
$$1 + 3 - 1 + 0$$

$$3 \cdot -1 = -3$$
$$3 - 3 = 0$$

3: Add the numbers in the dashed rectangle. Remember in math, the word "add" means addition or subtraction depending on the signs
4: Write the answer of step 3 in this position

Remember from the first step where our answer will be showing. One more thing, the synthetic divisions gives us the answer in terms of coefficients:

$$Coefficients\ of\ the\ answer = 1 + 3 - 1$$
$$Our\ reduced\ answer = 1x^2 + 3x - 1$$

Hint: If the last number of the synthetic division is NOT zero, then we did something wrong
Same as the polynomial division, the final reduced solution from which we can solve Is the multiplication of the synthetic answer and the initial zero:

Our initial equation had a highest exponent of x^3 therefore our reduced equation MUST start with x^2

$$(x - 3)(x^2 + 3x - 1)$$

Number in red is the operation performed in the current step
Number in blue is the operation performed in the previous step

In order to complete the solution, we have to set the equation equal to 0 as it originally was written

$$(x - 3)(x^2 + 3x - 1) = 0$$

Apply the zero product property

$x - 3 = 0$ | $x^2 + 3x - 1 = 0$

$x - 3 = 0$

$1x^2 + 3x - 1 = 0$

Moves as positive

$A = 1 \, ; B = +3 \, ; C = -1$

Identify the values of A, B and C

$x = 0 + 3$

$$x = \frac{-(3) \pm \sqrt{(3)^2 - 4 \cdot (1) \cdot (-1)}}{2 \cdot (1)}$$

$x = 3$

$$x = \frac{-3 \pm \sqrt{9 + 4}}{2}$$

$$x = \frac{-3 \pm \sqrt{13}}{2}$$

$$x = \frac{-3 + \sqrt{13}}{2} \quad ; \quad x = \frac{-3 - \sqrt{13}}{2}$$

$$x = 3 \, ; x = \frac{-3 + \sqrt{13}}{2} \text{ and } x = \frac{-3 - \sqrt{13}}{2} \quad \text{»» Answers}$$

Just one more big copy and paste, as we said, this last step is the same for the polynomial and the synthetic division

Number in red is the operation performed in the current step
Number in blue is the operation performed in the previous step

Problem 74) $x^4 + 2x^3 - 7x^2 - 8x + 12 = 0$

$$x^4 + 2x^3 - 7x^2 - 8x + 12 = 0$$

Find the initial zero

$$1 \cdot x^4 + 2x^3 - 7x^2 - 8x + 12 = 0$$

Multiply the first and last numbers

$$1 \cdot 12 = 12$$

Find all factors of the answer

$$Factors\ of\ 12 = \pm1;\ \pm2; \pm3;\ \pm4; \pm6; \pm12$$

Try each factor on the equation until it gives an answer of 0

$$For\ this\ equation, x = 1\ is\ a\ zero$$

$$1(1)^4 + 2(1)^3 - 7(1)^2 - 8(1) + 12$$

$$1(1) + 2(1) - 7(1) - 8(1) + 12$$

$$1 + 2 - 7 - 8 + 12 = 0$$

$$Yes, x = 2\ is\ a\ zero!$$

Now that we have our initial zero, we must set up

the: <u>Polynomial Division:</u>

Again, we already knew what the zero was, I mean, we invented the problem! But you should give it a try on your own

Number in red is the operation performed in the current step
Number in blue is the operation performed in the previous step

1st step:

1: Here we place the value we found as or initial zero. Very important to change its sign.
2: This is where our answer will be showing

$$x - 1 \overline{\smash{\big)}\ x^4 + 2x^3 - 7x^2 - 8x + 12}$$

3: In this position we place the equation we are trying to reduce

2nd step:

1: Divide the element in the blue square by the element in the thick red one:
$$\frac{x^4}{x} = x^3$$
2: Place the answer of step 1 here

$$x^3$$

$$\boxed{x} - 1 \overline{\smash{\big)}\ \boxed{x^4} + 2x^3 - 7x^2 - 8x + 12}$$
$$x^4 - x^3$$
$$3x^3 - 7x^2$$

3: Multiply the answer of step 1 by the x-1 at the "zero" location
4: Place the answer of step 3 in this position
5: Subtract elements in the dashed rectangle
6: Place the answer of step 5 in this position
7: Move the number in red to the position with the number in blue

$$(x - 1) \cdot x^3 = x \cdot x^3 - 1 \cdot x^3 = x^4 - x^3$$

$$(x^4 + 2x^3) - (x^4 - x^3) = x^4 + 2x^3 - x^4 + x^3$$
$$= x^4 - x^4 + 2x^3 + x^3 = 3x^2$$

Number in red is the operation performed in the current step
Number in blue is the operation performed in the previous step

108

3rd step:

1: Divide the element in the blue square by the element in the thick red one:
$$\frac{3x^3}{x} = 3x^2$$
2: Place the answer of step 1 here

3: Multiply the answer of step 1 by the x-1 at the "zero" location
4: Place the answer of step 3 in this position
5: Subtract elements in the dashed rectangle
6: Place the answer of step 5 in this position
7: Move the number in red to the position with the number in blue

$$x^3 + 3x^2 \quad \boxed{02}$$

$$\boxed{x} - 1 \quad \overline{\smash{\big)}\, x^4 + 2x^3 - 7x^2 - 8x + 12}$$

$$x^4 - x^3$$

$$\boxed{3x^3} - 7x^2$$

$$3x^3 - 3x^2$$

$$-4x^2 - 8x$$

$$(x-1) \cdot 3x^2 = x \cdot 3x^2 - 1 \cdot 3x^2 = 3x^3 - 3x^2$$

$$(3x^3 - 7x^2) - (3x^3 - 3x^2) = 3x^3 - 7x^2 - 3x^3 + 3x^2$$
$$= 3x^3 - 3x^3 - 7x^2 + 3x^2 = -4x^2$$

4th step:

1: Divide the element in the blue square by the element in the thick red one:
$$\frac{-4x^2}{x} = -4x$$
2: Place the answer of step 1 here

3: Multiply the answer of step 1 by the x-1 at the "zero" location
4: Place the answer of step 3 in this position
5: Subtract elements in the dashed rectangle
6: Place the answer of step 5 in this position
7: Move the number in red to the position with the number in blue

$$x^3 + 3x^2 - 4x \quad \boxed{02}$$

$$\boxed{x} - 1 \quad \overline{\smash{\big)}\, x^4 + 2x^3 - 7x^2 - 8x + 12}$$

$$x^4 - x^3$$

$$3x^3 - 7x^2$$

$$3x^3 - 3x^2$$

$$\boxed{-4x^2} - 8x$$

$$-4x^2 + 4x$$

$$-12x + 12$$

$$(x-1) \cdot -4x = x \cdot -4x - 1 \cdot -4x = -4x^4 + 4x$$

$$(-4x^2 - 8x) - (-4x^3 + 4x) = -4x^2 - 8x + 4x^2 - 4x$$
$$= -4x^2 + 4x^2 - 8x - 4x = -12x$$

Number in red is the operation performed in the current step
Number in blue is the operation performed in the previous step

109

5th step:

1: Divide the element in the blue square by the element in the thick red one:
$$\frac{-12x}{x} = -12$$

2: Place the answer of step 1 here

$$
\begin{array}{r}
x^3+3x^2-4x-12 \\
x-1 \enclose{longdiv}{x^4 + 2x^3 - 7x^2 - 8x + 12} \\
x^4 - x^3 \\
3x^3 - 7x^2 \\
3x^3 - 3x^2 \\
-4x^2 - 8x \\
-4x^2 + 4x \\
-12x + 12 \\
-12x + 12 \\
0 + 0
\end{array}
$$

3: Multiply the answer of step 1 by the x-1 at the "zero" location
4: Place the answer of step 3 in this position
5: Subtract elements in the dashed rectangle
6: Place the answer of step 5 in this position
7: Our final subtraction is equal to 0 and there are no more numbers to bring down. The division is over

$$(x-1) \cdot 12 = x \cdot -12 - 1 \cdot -12 = -12x + 12$$

$$(-12x + 12) - (-12x + 12) = -12x + 12 + 12x - 12$$
$$= -12x + 12x + 12 - 12 = 0 + 0$$

Remember the first step of the division. Our answer is showing at the top of the division

Our reduced equation is the multiplication of our zero and the answer of the division:

$$x^4 + 2x^3 - 7x^2 - 8x + 12 \rightarrow Reduced\ to \rightarrow (x-1)(x^3 + 3x^2 - 4x - 12)$$

Number in red is the operation performed in the current step
Number in blue is the operation performed in the previous step

110

Our new equation is now:

$$(x - 1)(x^3 + 3x^2 - 4x - 12) = 0$$

Apply the zero product property

$x - 1 = 0$ | $x^3 + 3x^2 - 4x - 12 = 0$

$x - 1 = 0$ $x^3 + 3x^2 - 4x - 12 = 0$

Moves as positive

$x = 0 + 1$

$x = 1$

Remember this answer for the end

This side requires another polynomial division as the highest power of X is bigger than 2

Number in red is the operation performed in the current step

Number in blue is the operation performed in the previous step

<u>2nd polynomial division</u>

$$x^3 + 3x^2 - 4x - 12 = 0$$

> Find the initial zero

$$1x^3 - 5x^2 - 4x - 12 = 0$$

> Multiply the first and last numbers

$$1 \cdot 12 = 12$$

> Find all factors of the answer

Yeap, signs don't really matter, we used a positive 12 even though it was negative

$$Factors\ of\ 12 = \pm1;\ \pm2; \pm3;\ \pm4; \pm6; \pm12$$

> Try each factor on the equation until it gives an answer of 0

$$For\ this\ equation, x = -3\ is\ a\ zero$$

$$(-3)^3 + 3(-3)^2 - 4(-3) - 12$$

$$(-27) + 3(9) - 4(-3) - 12$$

$$-27 + 27 + 12 - 12 = 0$$

Yes, x = -3 is a zero!

Now that we have our initial zero, we must set up the:
<u>Polynomial Division:</u>

$$x + 3 \enclose{longdiv}{x^3 + 3x^2 - 4x - 12}$$

1: Here we place the value we found as or initial zero. Very important to change its sign.
2: This is where our answer will be showing

3: In this position we place the equation we are trying to reduce

Number in red is the operation performed in the current step
Number in blue is the operation performed in the previous step

112

2nd step:

1: Divide the element in the blue square by the element in the thick red one:
$$\frac{x^3}{x} = x^2$$

2: Place the answer of step 1 here

$$x^2 \quad \boxed{02}$$

$$\boxed{x} + 3 \quad \overline{\left| x^3 \right| + 3x^2 - 4x - 12}$$

$$x^3 + 3x^2$$

$$0 - 4x - 12$$

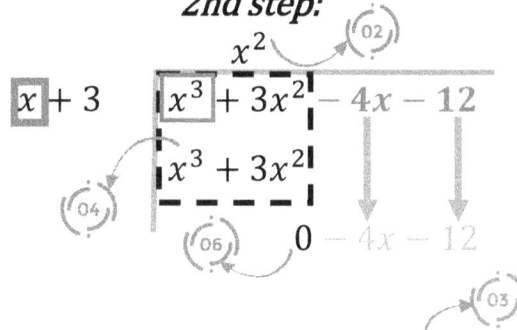

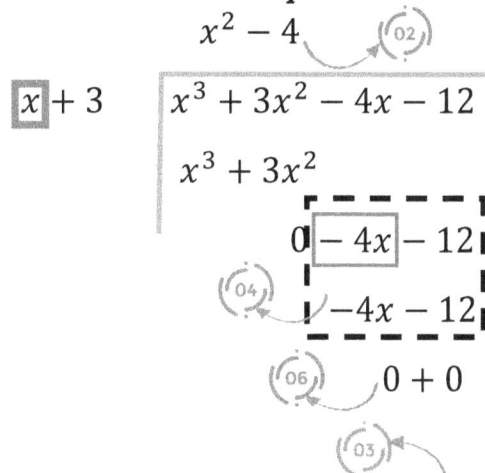

3: Multiply the answer of step 1 by the x+3 at the "zero" location
4: Place the answer of step 3 in this position
5: Subtract elements in the dashed rectangle
6: Place the answer of step 5 in this position
7: Since we got 0 in step 6, bring down TWO numbers in red to the blue position

$$(x + 3) \cdot x^2 = x \cdot x^2 + 3 \cdot x^2 = x^3 + 3x^2$$

$$(x^3 + 3x^2) - (x^3 + 3x^2) = x^3 + 3x^2 - x^3 - 3x^2$$
$$= x^3 - x^3 + 3x^2 - 3x^2 = 0$$

3rd step:

1: Divide the element in the blue square by the element in the thick red one:
$$\frac{-4x}{x} = -4$$

2: Place the answer of step 1 here

$$x^2 - 4 \quad \boxed{02}$$

$$\boxed{x} + 3 \quad \overline{\left| x^3 + 3x^2 - 4x - 12 \right.}$$

$$x^3 + 3x^2$$

$$0 - 4x - 12$$
$$-4x - 12$$

$$0 + 0$$

3: Multiply the answer of step 1 by the x+3 at the "zero" location
4: Place the answer of step 3 in this position
5: Subtract elements in the dashed rectangle
6: Place the answer of step 5 in this position
7: Our final subtraction is equal to 0 and there are no more numbers to bring down. The division is over

$$(x + 3) \cdot -4 = x \cdot -4 + 3 \cdot -4 = -4x - 12$$

$$(-4x - 12) - (-4x - 12) = -4x - 12 + 4x + 12$$
$$= -4x + 4x - 12 + 12 = 6x$$

Remember the first step of the division. Our answer is showing at the top of the division
Our reduced equation is the multiplication of our zero and the answer of the division:

$$x^3 + 3x^2 - 4x - 12 \rightarrow Reduced\ to \rightarrow (x + 3)(x^2 - 4)$$

Number in red is the operation performed in the current step
Number in blue is the operation performed in the previous step

In order to complete the solution, we have to set the equation equal to 0 as it originally was written

$$(x + 3)(x^2 - 4) = 0$$

Apply the zero product property

$x + 3 = 0$ | $x^2 - 4 = 0$

$x + 3 = 0$ | $x^2 - 4 = 0$

Moves as negative | Moves as positive

$x = 0 - 3$ | $x^2 = 0 + 4$

$x = 3$

Moves as a root

$$x = \sqrt[2]{4}$$

$$x = \pm 2$$

$$x = 1; x = -3 ; x = 2 \text{ and } x = -2 \text{ »» Answers}$$

Do not forget about this answer obtained at the first division

Problem 74-b) $x^4 + 2x^3 - 7x^2 - 8x + 12 = 0$

$$x^4 + 2x^3 - 7x^2 - 8x + 12 = 0$$

<u>Synthetic division:</u>

Keep in mind from previous steps we already know the following:

$$x = 1 \ is \ a \ zero$$

The **1st step** is to set up the division:

1: This is the position where we place our zero. For the synthetic division we don't change its sign
2: This is where our answer will be showing

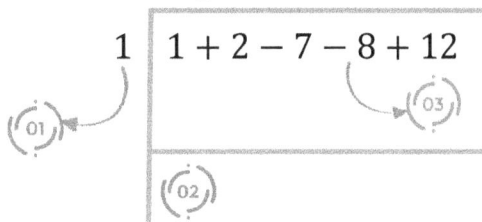

3: Here we will place the coefficients of the equation we are trying to reduce:

For the equation:

$$1x^4 + 2x^3 - 7x^2 - 8x + 12$$

Its coefficients are:

$$1 + 2 \ - 7 - 8 + 12$$

2nd step:

1: Bring down the number colored in red to the position of the number in blue
2: Multiply the number in the thick red box with the number in the blue box
3: Place the answer of step two in this position

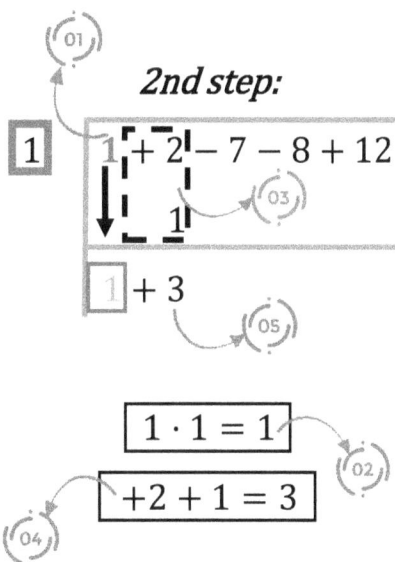

4: Add the numbers in the dashed rectangle. Remember in math, the word "add" means addition or subtraction depending on the signs
5: Write the answer of step 4 in this position.

Number in red is the operation performed in the current step
Number in blue is the operation performed in the previous step

3rd step:

1: Multiply the number in the thick red box with the number in the blue box
2: Place the answer of step one in this position

1 $1 + 2 - 7 - 8 + 12$

$1 + 3$ (02)

$1 + 3 - 4$ (04)

$1 \cdot 3 = 3$ (01)

$-7 + 3 = -4$ (03)

3: Add the numbers in the dashed rectangle. Remember in math, the word "add" means addition or subtraction depending on the signs
4: Write the answer of step 3 in this position

4th step:

1: Multiply the number in the thick red box with the number in the blue box
2: Place the answer of step one in this position

1 $1 + 2 - 7 - 8 + 12$

$1 + 3 - 4$ (02)

$1 + 3 - 4 - 12$ (04)

$1 \cdot -4 = -4$ (01)

$-8 - 4 = -12$ (03)

3: Add the numbers in the dashed rectangle. Remember in math, the word "add" means addition or subtraction depending on the signs
4: Write the answer of step 3 in this position

5th step:

1 $1 + 2 - 7 - 8 + 12$

$1 + 3 - 4 - 12$ (02)

$1 + 3 - 4 - 12 + 0$ (04)

1: Multiply the number in the thick red box with the number in the blue box
2: Place the answer of step one in this position

$1 \cdot -12 = -12$ (01)

$12 - 12 = 0$ (03)

3: Add the numbers in the dashed rectangle. Remember in math, the word "add" means addition or subtraction depending on the signs
4: Write the answer of step 3 in this position

Number in red is the operation performed in the current step
Number in blue is the operation performed in the previous step

Remember from the first step where our answer will be showing. One more thing, the synthetic divisions gives us the answer in terms of coefficients:

$$Coefficients\ of\ the\ answer = 1 + 3 - 4 - 12$$
$$Our\ reduced\ answer = 1x^3 + 3x^2 - 4x - 12$$

Hint: If the last number of the synthetic division is NOT zero, then we did something wrong

Same as the polynomial division,
The answer is the multiplication of the synthetic answer and the initial zero:

Our initial equation had a highest exponent of x^4 therefore our reduced equation MUST start with x^3

$$(x - 1)(x^3 + 3x^4 - 4x - 12)$$

Our new equation is now:

$$(x - 1)(x^3 + 3x^2 - 4x - 12) = 0$$

Apply the zero product property

$x - 1 = 0$

$x^3 + 3x^2 - 4x - 12 = 0$

$x - 1 = 0$

$$x^3 + 3x^2 - 4x - 12 = 0$$

Moves as positive

$x = 0 + 1$

$x = 1$

Remember this answer for the end

This side requires another polynomial division as the highest power of x is bigger than 2

Number in red is the operation performed in the current step
Number in blue is the operation performed in the previous step

117

<u>2nd Synthetic division:</u>

$$x^3 + 3x^2 - 4x - 12 = 0$$

Remember from the solutions of the polynomial division that:

$$x = -3 \; is \; a \; zero$$

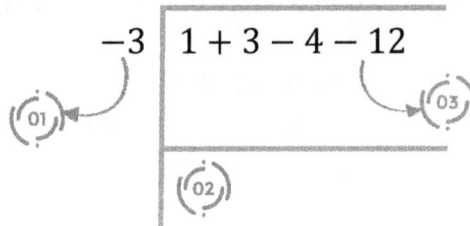

1: This is the position where we place our zero. For the synthetic division we don't change its sign
2: This is where our answer will be showing

3: Here we will place the coefficients of the equation we are trying to reduce:

For the equation:

$$1x^3 + 3x^2 - 4x - 12$$

Its coefficients are:

$$1 + 3 - 4 - 12$$

2nd step:

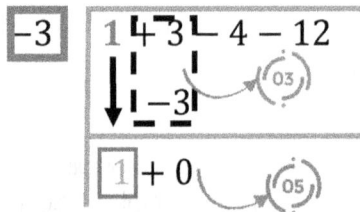

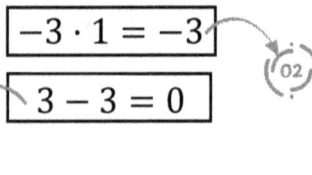

1: Bring down the number colored in red to the position of the number in blue
2: Multiply the number in the thick red box with the number in the blue box
3: Place the answer of step two in this position

4: Add the numbers in the dashed rectangle. Remember in math, the word "add" means addition or subtraction depending on the signs
5: Write the answer of step 4 in this position.

Number in red is the operation performed in the current step
Number in blue is the operation performed in the previous step

118

3rd step:

1: Multiply the number in the thick red box with the number in the blue box
2: Place the answer of step one in this position

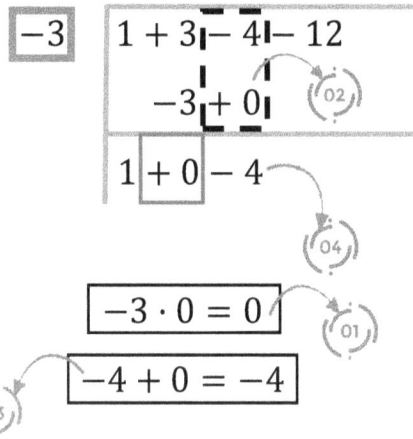

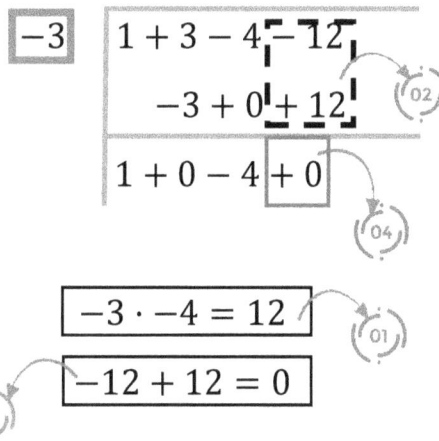

$$\boxed{-3} \quad \begin{array}{r} 1 + 3 \;\lceil - 4 \rceil - 12 \\ -3 \lceil + 0 \rceil \\ \hline 1 \;\boxed{+ 0} - 4 \end{array}$$

3: Add the numbers in the dashed rectangle. Remember in math, the word "add" means addition or subtraction depending on the signs
4: Write the answer of step 3 in this position

$$\boxed{-3 \cdot 0 = 0}$$
$$\boxed{-4 + 0 = -4}$$

4th step:

1: Multiply the number in the thick red box with the number in the blue box
2: Place the answer of step one in this position

$$\boxed{-3} \quad \begin{array}{r} 1 + 3 - 4 \;\lceil - 12 \rceil \\ -3 + 0 \lceil + 12 \rceil \\ \hline 1 + 0 - 4 \;\boxed{+ 0} \end{array}$$

3: Add the numbers in the dashed rectangle. Remember in math, the word "add" means addition or subtraction depending on the signs
4: Write the answer of step 3 in this position

$$\boxed{-3 \cdot -4 = 12}$$
$$\boxed{-12 + 12 = 0}$$

Remember from the first step where our answer will be showing. One more thing, the synthetic divisions gives us the answer in terms of coefficients:

$$Coefficients\ of\ the\ answer = 1 + 0 - 4$$

$$Our\ reduced\ answer = 1x^2 + 0x - 4$$

Our initial equation had a highest exponent of x^3 therefore our reduced equation MUST start with x^2

Hint: If the last number of the synthetic division is NOT zero,

then we did something wrong

Same as the polynomial division,

The answer is the multiplication of the synthetic answer and the initial zero:

$$(x + 3)(x^2 - 4)$$

Number in red is the operation performed in the current step
Number in blue is the operation performed in the previous step

In order to complete the solution, we have to set the equation equal to 0 as it originally was written

$$(x + 3)(x^2 - 4) = 0$$

Apply the zero product property

$x + 3 = 0$ | $x^2 - 4 = 0$

$x + 3 = 0$ | $x^2 - 4 = 0$

Moves as negative | Moves as positive

$x = 0 \; ^{-3}$ | $x^2 = 0 \; + 4$

| Moves as a root

$x = 3$

$x = \sqrt[2]{4}$

$x = \pm 2$

$$x = 1; x = -3; x = 2 \text{ and } x = -2 \text{ »» Answers}$$

Do not forget about this answer obtained at the first division

Higher degree polynomials: practice problems

1) $x^3 - 2x^2 - 13x - 10 = 0$ 2) $x^3 - 19x^2 + 108x - 180 = 0$

3) $x^3 + 2x^2 - x - 2 = 0$ 4) $2x^3 + x^2 - 13x + 6 = 0$

5) $x^4 - 15x^2 + 10x + 24 = 0$ 6) $x^4 - 4x^3 - 9x^2 + 16x + 20 =$

Number in red is the operation performed in the current step
Number in blue is the operation performed in the previous step

The good news is, we finished the hardest section of this book; that higher degree polynomial section we must admit it was painful. So far we've learned how to solve problems with many different types of numbers around x; some of them multiplying, some adding, subtracting, dividing, we have even learned how to solve with different exponents and roots. The techniques explained up to this point are enough to solve a wide variety of problems but, what happens when x becomes the exponent?

$$2^x = 16$$

None of the methods detailed before will be useful to solve this problem. Think about it, that x is on the exponent, if we apply the opposite's rule, we would move it to the other side as a root:

$$2^x = 16$$

Moves as a root

$$2 = \sqrt{16}$$

In any case, we just made the problem harder, now we are trying to find the root x of 16... wait what?!

In this section we will learn how to solve these problems using the similar bases method and logarithms. As usual we will use detailed examples to explain each method with different scenarios but before we jump into them, we must clarify a few rules of exponents that will come handy:

Multiplication of equal bases = keep the base, add the exponents

$$A^M \cdot A^N = A^{M+N} \rightarrow Example \rightarrow 3^4 \cdot 3^5 = 3^{4+5} = 3^9$$

Division of equal bases = keep the bases, subtract the exponents.

Important note: The exponent of the numerator goes first in the subtraction

$$\frac{A^M}{A^N} = A^{M-N} \rightarrow Example \rightarrow \frac{5^7}{5^2} = 5^{7-2} = 5^5$$

The exponent of an exponent = multiply the exponents together, keep the base

$$(A^M)^N = A^{M \cdot N} \rightarrow Example \rightarrow (2^4)^6 = 2^{4 \cdot 6} = 2^{24}$$

Roots = a fractionated exponent

$$\sqrt[N]{A^M} = A^{M/N} \rightarrow Example \rightarrow \sqrt[3]{2^4} = 2^{4/3}$$

Number in red is the operation performed in the current step
Number in blue is the operation performed in the previous step

121

Negative exponent = the exponent is in the wrong place

$$A^{-M} = \frac{1}{A^M} \rightarrow Example \rightarrow 2^{-4} = \frac{1}{2^4}$$

Also

$$\frac{1}{A^{-M}} = A^M \rightarrow Example \rightarrow \frac{1}{3^{-5}} = 3^5$$

Similar bases method

The easiest way of solving problems involving x's in the exponents position is to force every element in the equation to have the same base, then simply eliminate the bases and be left with the exponents:

$$4^x = 16$$

Turn the 16 into a base of 4

$$4^x = 4^2$$

Remember we can do anything to the equation as long as we do it to the other side:
Eliminate the base of 4 on both sides

$$x = 2$$

Let's view some detailed examples so we can get more experience with this method.

Number in red is the operation performed in the current step
Number in blue is the operation performed in the previous step

122

Problem 75) $3^x = 81$

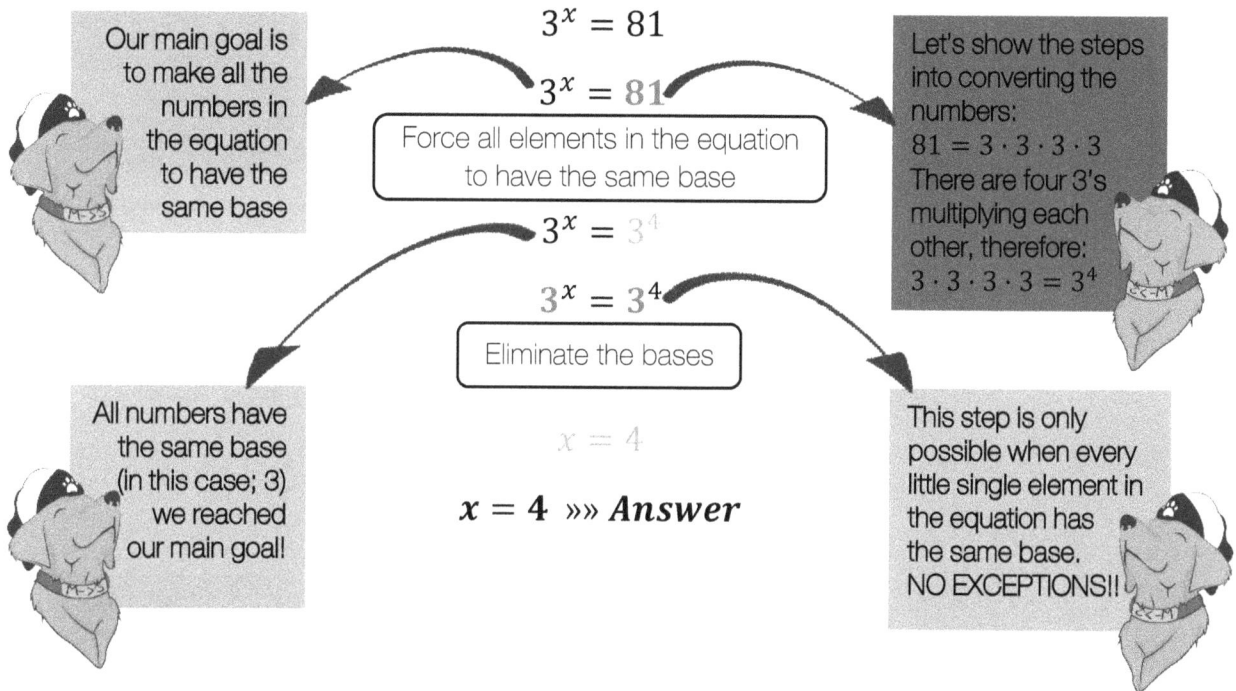

$$3^x = 81$$

Our main goal is to make all the numbers in the equation to have the same base

$$3^x = 81$$

Force all elements in the equation to have the same base

$$3^x = 3^4$$

$$3^x = 3^4$$

Eliminate the bases

$$x = 4$$

$$x = 4 \text{ »» } \textbf{\textit{Answer}}$$

Let's show the steps into converting the numbers:
$81 = 3 \cdot 3 \cdot 3 \cdot 3$
There are four 3's multiplying each other, therefore:
$3 \cdot 3 \cdot 3 \cdot 3 = 3^4$

All numbers have the same base (in this case; 3) we reached our main goal!

This step is only possible when every little single element in the equation has the same base. NO EXCEPTIONS!!

Problem 76) $5^x = 125$

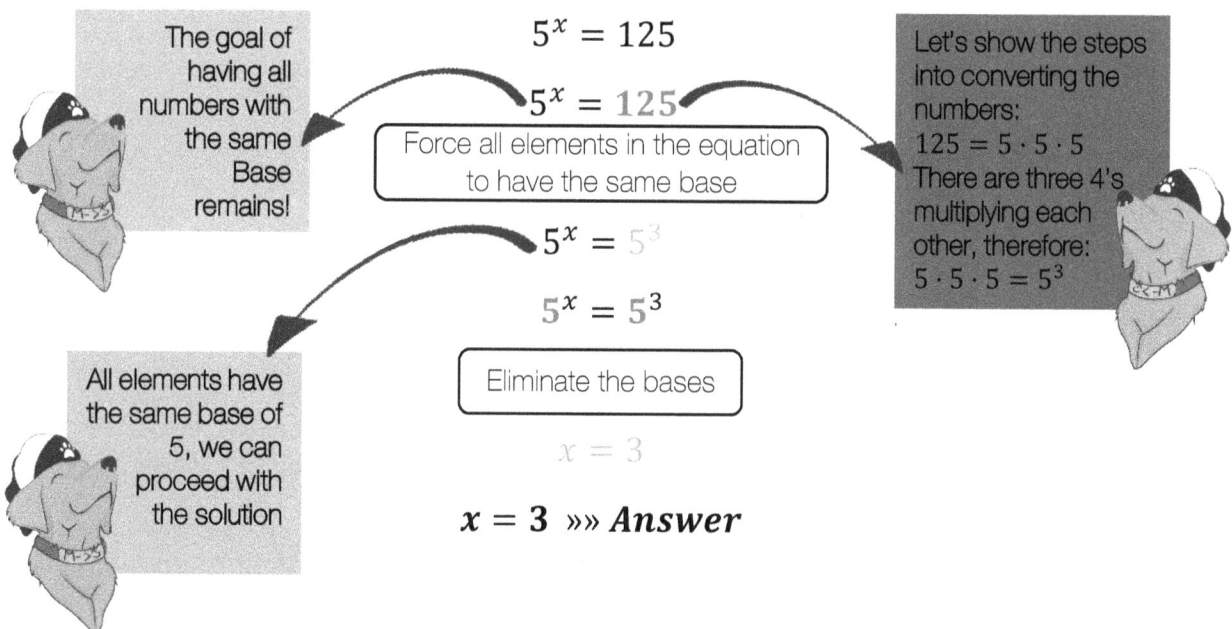

$$5^x = 125$$

The goal of having all numbers with the same Base remains!

$$5^x = 125$$

Force all elements in the equation to have the same base

$$5^x = 5^3$$

$$5^x = 5^3$$

Eliminate the bases

$$x = 3$$

$$x = 3 \text{ »» } \textbf{\textit{Answer}}$$

Let's show the steps into converting the numbers:
$125 = 5 \cdot 5 \cdot 5$
There are three 4's multiplying each other, therefore:
$5 \cdot 5 \cdot 5 = 5^3$

All elements have the same base of 5, we can proceed with the solution

Number in red is the operation performed in the current step
Number in blue is the operation performed in the previous step

Problem 77) $6^{3x} \cdot 6^x = 36$

$$6^{3x} \cdot 6^x = 36$$

$$6^{3x} \cdot 6^x = 36$$

Remember that thing about putting all elements with x together? Well, it still applies

$$\boxed{Apply\ A^M \cdot A^N = A^{M+N}}$$

$$6^{3x+x} = 36$$

$$6^{4x} = 36$$

$$\boxed{\text{Force all elements in the equation to have the same base}}$$

One more time:
$36 = 6 \cdot 6$
There are two 6's:
$6 \cdot 6 = 6^2$

$$6^{4x} = 6^2$$

$$6^{4x} = 6^2$$

$$\boxed{\text{Eliminate the bases}}$$

$$4x = 2$$

$$4x = 2$$

$$\boxed{\text{Moves as division}}$$

Don't forget to simplify all fractions as much as possible:

$$\frac{2 \div 2}{4 \div 2} = \frac{1}{2}$$

$$x = \frac{2}{4}$$

$$x = \frac{1}{2} \quad \text{»» } \textbf{Answer}$$

Number in red is the operation performed in the current step
Number in blue is the operation performed in the previous step

124

Problem 78) $4^x \cdot 64^{2x} = 1024$

We really want all elements to have the same base, is literally the name of the method

$$4^x \cdot 64^{2x} = 1024$$

$$4^x \cdot 64^{2x} = 1024$$

Force all elements in the equation to have the same base

$$4^x \cdot (4^3)^{2x} = 4^5$$

Double time!

$$64 = 4 \cdot 4 \cdot 4 = 4^3$$

And the other one:

$$1024 = 4 \cdot 4 \cdot 4 \cdot 4 \cdot 4 = 4^5$$

$$4^x \cdot (4^3)^{2x} = 4^5$$

Apply $(A^M)^N = A^{M \cdot N}$

That parenthesis is an obstacle, good thing we have a rule to get rid of it

$$4^x \cdot 4^{3 \cdot 2x} = 4^5$$

$$4^x \cdot 4^{6x} = 4^5$$

Apply $A^M \cdot A^N = A^{M+N}$

One more time, put all the elements with x together!

$$4^{x+6x} = 4^5$$

$$4^{7x} = 4^5$$

Eliminate the bases

$$7x = 5$$

$$7x = 5$$

Moves as division

$$x = \frac{5}{7}$$

$$x = \frac{5}{7} \quad \text{»» \textbf{Answer}}$$

Number in red is the operation performed in the current step
Number in blue is the operation performed in the previous step

125

Problem 79) $49^x \cdot \sqrt{7^x} = 343$

$$49^x \cdot \sqrt{7^x} = 343$$

Remember the name of the method; SIMILAR bases

$$49^x \cdot \sqrt{7^x} = 343$$

Force all elements in the equation to have the same base

Double time!
$49 = 7 \cdot 7 = 7^2$
And the other one:
$343 = 7 \cdot 7 = 7^3$

$$(7^2)^x \cdot \sqrt{7^x} = 7^3$$

$$(7^2)^x \cdot \sqrt{7^x} = 7^3$$

Apply $(A^M)^N = A^{M \cdot N}$

Recall:

$$\sqrt{7^x} = \sqrt[2]{7^x}$$

$$7^{2 \cdot x} \cdot \sqrt{7^x} = 7^3$$

$$7^{2x} \cdot \sqrt{7^x} = 7^3$$

Apply $\sqrt[N]{A^M} = A^{M/N}$

We are just trying to make all elements to look smooth; no roots, no parenthesis, nothing fancy

$$7^{2x} \cdot 7^{\frac{x}{2}} = 7^3$$

$$7^{2x} \cdot 7^{\frac{x}{2}} = 7^3$$

Apply $A^M \cdot A^N = A^{M+N}$

For addition of fractions apply:
$$\frac{A}{B} + \frac{C}{D} = \frac{A \cdot D + B \cdot C}{B \cdot D}$$

$$\frac{2x}{1} + \frac{x}{2} = \frac{2x \cdot 2 + 1 \cdot x}{1 \cdot 2}$$

$$= \frac{4x + x}{2} = \frac{5x}{2}$$

$$7^{2x + \frac{x}{2}} = 7^3$$

$$7^{\frac{5x}{2}} = 7^3$$

Eliminate the bases

$$\frac{5x}{2} = 3$$

$$\frac{5x}{2} = 3$$

Moves as multiplication

$$5x = 3 \cdot 2$$

Moves as division

$$x = \frac{6}{5} \;\text{»»}\; x = \frac{6}{5} \;\text{»» } \boldsymbol{Answer}$$

Number in red is the operation performed in the current step
Number in blue is the operation performed in the previous step

Problem 80) $3^{3x} \cdot \sqrt[5]{3^{2x}} = 243$

$$3^{3x} \cdot \sqrt[5]{3^{2x}} = 243$$

$$3^{3x} \cdot \sqrt[5]{3^{2x}} = 243$$

We'll keep helping you with this part:
$243 = 3 \cdot 3 \cdot 3 \cdot 3 \cdot 3$
$= 3^5$

Force all elements in the equation to have the same base

$$3^{3x} \cdot \sqrt[5]{3^{2x}} = 3^5$$

$$3^{3x} \cdot \sqrt[5]{3^{2x}} = 3^5$$

Apply $\sqrt[N]{A^M} = A^{M/N}$

$$3^{3x} \cdot 3^{\frac{2x}{5}} = 3^5$$

$$3^{3x} \cdot 3^{\frac{2x}{5}} = 3^5$$

Apply $A^M \cdot A^N = A^{M+N}$

For addition of fractions apply:

$$\frac{A}{B} + \frac{C}{D} = \frac{A \cdot D + B \cdot C}{B \cdot D}$$

$$\frac{3x}{1} + \frac{2x}{5} = \frac{3x \cdot 5 + 1 \cdot 2x}{1 \cdot 5}$$

$$= \frac{15x + 2x}{5} = \frac{17x}{5}$$

$$3^{3x + \frac{2x}{5}} = 3^5$$

$$3^{\frac{17x}{5}} = 3^5$$

Eliminate the bases

$$\frac{17x}{5} = 5$$

$$\frac{17x}{5} = 5$$

Moves as multiplication

$$17x = 5 \cdot 5$$

Moves as division

$$x = \frac{25}{17}$$

$$x = \frac{25}{17} \text{ »» } \textit{Answer}$$

Number in red is the operation performed in the current step
Number in blue is the operation performed in the previous step

Problem 81) $\frac{2^{9x} \cdot 8^{3x}}{16^{4x}} = 256$

$$\frac{2^{9x} \cdot 8^{3x}}{16^{4x}} = 256$$

$$\frac{2^{9x} \cdot 8^{3x}}{16^{4x}} = 256$$

Force all elements in the equation
to have the same base

$$\frac{2^{9x} \cdot (2^3)^{3x}}{(2^4)^{4x}} = 2^8$$

No worries, we know is hard
sometimes, we'll keep helping
you:
$256 = 2 \cdot 2 \cdot 2 \cdot 2 \cdot 2 \cdot 2 \cdot 2 \cdot 2$
$= 2^8$
$16 = 2 \cdot 2 \cdot 2 \cdot 2 = 2^4$
$8 = 2 \cdot 2 \cdot 2 = 2^3$

$$\frac{2^{9x} \cdot (2^3)^{3x}}{(2^4)^{4x}} = 2^8$$

Apply $(A^M)^N = A^{M \cdot N}$

$$\frac{2^{9x} \cdot 2^{3 \cdot 3x}}{2^{4 \cdot 4x}} = 2^8$$

$$\frac{2^{9x} \cdot 2^{9x}}{2^{16x}} = 2^8$$

Apply $A^M \cdot A^N = A^{M+N}$

$$\frac{2^{9x+9x}}{2^{16x}} = 2^8$$

$$\frac{2^{18x}}{2^{16x}} = 2^8$$

Apply $\frac{A^M}{A^N} = A^{M-N}$

All that process
just to get some
nice similar
bases we can
get rid of

$$2^{18x-16x} = 2^8$$

$$2^{2x} = 2^8$$

Eliminate the bases

$$2x = 8$$

Number in red is the operation performed in the current step
Number in blue is the operation performed in the previous step

$$2x = 8$$

Moves as division

$$x = \frac{8}{2}$$

$$x = 4 \; »» \; Answer$$

Problem 82) $\frac{5^{3x}}{25^{4x}} = 5^4$

$$\frac{5^{3x}}{25^{4x}} = 5^4$$

$$\frac{5^{3x}}{25^{4x}} = 5^4$$

We said we got you on this!

$$25 = 5 \cdot 5 = 5^2$$

Force all elements in the equation to have the same base

$$\frac{5^{3x}}{(5^2)^{4x}} = 5^4$$

Apply $(A^M)^N = A^{M \cdot N}$

$$\frac{5^{3x}}{5^{2 \cdot 4x}} = 5^4$$

$$\frac{5^{3x}}{5^{8x}} = 5^4$$

Apply $\frac{A^M}{A^N} = A^{M-N}$

$$5^{3x-8x} = 5^4$$

$$5^{-5x} = 5^4$$

Just a quick addition/subtraction practice:
$$3x - 8x = -5x$$
Different signs mean subtraction. The 8 has the negative and is the highest number, means negative answer

Eliminate the bases

$$-5x = 4$$

$$-5x = 4$$

Moves as division

$$x = \frac{4}{-5} \; »» \; x = -\frac{4}{5} \; »» \; Answer$$

Number in red is the operation performed in the current step
Number in blue is the operation performed in the previous step

Problem 83) $\dfrac{2^{2x} \cdot 2^{4x}}{\sqrt[3]{2^{2x}}} = 2^{10}$

$$\dfrac{2^{2x} \cdot 2^{4x}}{\sqrt[3]{2^{2x}}} = 2^{10}$$

$$\dfrac{2^{2x} \cdot 2^{4x}}{\sqrt[3]{2^{2x}}} = 2^{10}$$

$$\boxed{Apply \ \sqrt[N]{A^M} = A^{M/N}}$$

It is very important to remember, every time we apply this rule the positions of the exponents. The exponent inside of the root is the numerator of the fraction and the exponent outside of the root is the denominator

$$\dfrac{2^{2x} \cdot 2^{4x}}{2^{\frac{2x}{3}}} = 2^{10}$$

$$\dfrac{2^{2x} \cdot 2^{4x}}{2^{\frac{2x}{3}}} = 2^{10}$$

$$\boxed{Apply \ A^M \cdot A^N = A^{M+N}}$$

$$\dfrac{2^{2x+4z}}{2^{\frac{2x}{3}}} = 2^{10}$$

$$\dfrac{2^{6x}}{2^{\frac{2x}{3}}} = 2^{10}$$

$$\boxed{Apply \ \dfrac{A^M}{A^N} = A^{M-N}}$$

Yeah, subtracting fractions is a bit hard sometimes. Just apply this method every time:
$$\dfrac{A}{B} - \dfrac{C}{D} = \dfrac{A \cdot D - B \cdot C}{B \cdot D}$$
$$\dfrac{6x}{1} - \dfrac{2x}{3} = \dfrac{6x \cdot 3 - 1 \cdot 2x}{1 \cdot 3}$$
$$= \dfrac{18x - 2x}{3} = \dfrac{16x}{3}$$

$$2^{6x - \frac{2x}{3}} = 2^{10}$$

$$2^{\frac{16x}{3}} = 2^{10}$$

$$\boxed{\text{Eliminate the bases}}$$

$$16x/3 = 10$$

Number in red is the operation performed in the current step
Number in blue is the operation performed in the previous step

$$\frac{16x}{3} = 10$$

Moves as multiplication

$$16x = 10 \cdot 3$$

Moves as division

$$x = \frac{30}{16}$$

Never forget to simplify your fractions!!
$$\frac{30 \div 2}{16 \div 2} = \frac{15}{8}$$

$$x = \frac{15}{8} \text{ »» } Answer$$

Problem 84) $\frac{1}{10^x} = 1000$

$$\frac{1}{10^x} = 1000$$

We are here for you!
$$1000 = 10 \cdot 10 \cdot 10$$
$$= 10^3$$

$$\frac{1}{10^x} = 1000$$

Force all elements in the equation to have the same base

$$\frac{1}{10^x} = 10^3$$

$$\frac{1}{10^x} = 10^3$$

Remember that thing about putting all elements with x together? Well, it still applies

$$Apply \; \frac{1}{A^M} = A^{-M}$$

$$10^{-x} = 10^3$$

$$10^{-x} = 10^3$$

Eliminate the bases

$$-x = 3$$

$$-x = 3$$

Move the negative to the other side

$$x = -3$$

$$x = -3 \text{ »» } Answer$$

Number in red is the operation performed in the current step

Number in blue is the operation performed in the previous step

Problem 85) $6^x + 6^x = 72$

$$6^x + 6^x = 72$$

$$6^x + 6^x = 72$$

$$2 \cdot 6^x = 72$$

Moves as division

Every time we have two or more elements EXACTLY equal to each other we can just add them in the same way we do with x. Think about it:

$x + x = 2x$

Therefore:

$6^x + 6^x = 2 \cdot 6^x$

$$6^x = \frac{72}{2}$$

$$6^x = 36$$

Force all elements in the equation to have the same base

One more time:

$36 = 6 \cdot 6 = 6^2$

$$6^x = 6^2$$

$$6^x = 6^2$$

Eliminate the bases

$$x = 2$$

$x = 2$ »» **Answer**

Number in red is the operation performed in the current step

Number in blue is the operation performed in the previous step

Problem 86) $7^{2x} \cdot 9^x = 49$

$$7^{2x} \cdot 9^x = 49$$

$$7^{2x} \cdot 9^x = 49$$

The bases of the exponents with x are not equal and there is no way we can actually make them equal. This problem requires more advanced math to be solved. This is beyond the scope of this book and most of your math classes, relax.

Problem 87) $4^x \cdot 24^x = 16$

$$4^x \cdot 24^x = 16$$

$$4^x \cdot 24^x = 16$$

The bases of the exponents with x are not equal and there is no way we can actually make them equal. This problem requires more advanced math to be solved. This is beyond the scope of this book and most of your math classes, relax. Yes, we did one more copy and paste in this book.

Problem 88) $2^x - 2^{3x} = 32$

$$2^x - 2^{3x} = 32$$

$$2^x - 2^{3x} = 32$$

Addition/subtraction problems can only be solved when both the base and the exponent are exactly the same, like problem 85. One more time, this problem is way beyond the scope of the book and most of your math lessons. Yes, last sentence was kind of a copy and paste.

Problem 89) $3^x + 5^x = 32$

$$3^x + 5^x = 32$$

$$3^x + 5^x = 32$$

Addition/subtraction problems can only be solved when both the base and the exponent are exactly the same, like problem 85. One more time, this problem is way beyond the scope of the book and most of your math lessons. Sorry, is the same answer, so we just copy and paste.

Number in red is the operation performed in the current step
Number in blue is the operation performed in the previous step

133

<u>Similar bases method: practice problems</u>

1) $8^x = 4096$ 2) $7^x = 343$

3) $\dfrac{1}{4^x} = 32$ 4) $3^x \cdot 81^x = 27$

5) $\dfrac{2^{8x} \cdot 4^{3x}}{16^x} = 256$ 6) $\dfrac{\sqrt{27^x} \cdot 3^{2x}}{9^{2x}} = 81$

Number in red is the operation performed in the current step

Number in blue is the operation performed in the previous step

Logarithms method

Following the analogy of math being a language, here is when we start with the crazy words that at first make no sense but as soon as we understand its meaning and how to properly use them, we start forcing them on every sentence possible just to look smart.

Through the entire length of this book we have been solving for x using the "opposites rule" to move things around the equations. Last method we learned (similar bases) wasn't really following that rule, it was more a "cancel this on one side and cancel the same thing on the other". With logarithms we come back to the opposites rule. Logarithms look like roots in one aspect, to solve most of them we will need a calculator. We will leave answers as exact answers, meaning we will not solve for the logarithm, but if you want to have a decimal answer, you will need a calculator.

In summary, logarithms are the opposite of exponents and we will use them to move things around the equation when x is at the exponent position. Check the following example:

$$2^x = 14$$

Moves as a logarithm

$$x = \log_2(14)$$

Take a look how the base of the exponent became the base of the logarithm. This is how we use logarithms to solve for equations involving x in the exponent. Same as the *similar bases method*, our first goal will be to simplify the equation until we are left with only one exponent with x, then, we apply the logarithm.

Problem 90) $7^x = 20$

$$7^x = 20$$

$$7^x = 20$$

Moves as a logarithm

$$x = \log_7(20)$$

We can't really apply the "similar bases" method for this problem as 20 can't be turned into a number with base of 7

$$x = \log_7(20) \; »» \; \textbf{\textit{Answer}}$$

This is what we meant with an exact answer. In order to get 1.5395 which is the decimal answer, you would need a calculator

Number in red is the operation performed in the current step
Number in blue is the operation performed in the previous step

Problem 91) $2^x = 17$

$$2^x = 17$$

$$2^x = 17$$

Moves as a logarithm

$$x = \log_2(17)$$

$$x = \log_2(17) \text{ »» } \textbf{\textit{Answer}}$$

Problem 92) $3^x \cdot 3^{2x} = 15$

$$3^x \cdot 3^{2x} = 15$$

$$3^x \cdot 3^{2x} = 15$$

Apply $A^M \cdot A^N = A^{M+N}$

$$3^{x+2x} = 15$$

$$3^{3x} = 15$$

Moves as a logarithm

$$3x = \log_3(15)$$

Moves as division

$$x = \frac{\log_3(15)}{3}$$

As usual, we first need to group all elements having x

$$x = \frac{\log_3(15)}{3} \text{ »» } \textbf{\textit{Answer}}$$

Problem 93) $4^{3x} \cdot 16^x = 35$

$$4^{3x} \cdot 16^x = 35$$

$$4^{3x} \cdot 16^x = 35$$

Force all elements with x in the exponent to have the same base

$$4^{3x} \cdot (4^2)^x = 35$$

You will not get lost in our watch!!

$16 = 4 \cdot 4 = 4^2$

This problem begins using the same technique as "similar bases" method. As we mentioned, we must first group all the x's together

Number in red is the operation performed in the current step
Number in blue is the operation performed in the previous step

All of these was just an attempt to have all the x's into only one element

$$4^{3x} \cdot \left(4^2\right)^x = 35$$

Apply $(A^M)^N = A^{M \cdot N}$

$$4^{3x} \cdot 4^{2 \cdot x} = 35$$

$$4^{3x} \cdot 4^{2x} = 35$$

Apply $A^M \cdot A^N = A^{M+N}$

$$4^{3x+2x} = 35$$

$$4^{5x} = 35$$

Moves as a logarithm

The main reason we are using logarithms is because 35 can NOT be turned into a base of 4

$$5x = \log_4(35)$$

Moves as division

$$x = \frac{\log_4(35)}{5}$$

$$x = \frac{\log_4(35)}{5} \quad »» \text{ Answer}$$

Problem 94) $4^x \cdot 24^x = 16$

$$4^x \cdot 24^x = 16$$

$$4^x \cdot 24^x = 16$$

Apply $A^M \cdot B^M = (A \cdot B)^M$

$$(4 \cdot 24)^x = 16$$

$$96^x = 16$$

Moves as a logarithm

This is a new rule. This time, the bases are different but the exponents are EXACTLY the same. In this situation we can just multiply the bases together and leave the same exponent. This is only possible if the exponents are EXACTLY equal

$$x = \log_{96}(16)$$

$$x = \log_{96}(16) \quad »» \text{ Answer}$$

Number in red is the operation performed in the current step
Number in blue is the operation performed in the previous step

Problem 95) $\frac{2^{2x} \cdot 16^{3x}}{8^{4x}} = 33$

$$\frac{2^{2x} \cdot 16^{3x}}{8^{4x}} = 33$$

$$\frac{2^{2x} \cdot 16^{3x}}{8^{4x}} = 33$$

> Force all elements with x in the exponent to have the same base

We are here for you:

$16 = 2 \cdot 2 \cdot 2 \cdot 2 = 2^4$
$8 = 2 \cdot 2 \cdot 2 = 2^3$

$$\frac{2^{2x} \cdot \left(2^4\right)^{3x}}{\left(2^3\right)^{3x}} = 33$$

$$\frac{2^{2x} \cdot \left(2^4\right)^{3x}}{\left(2^3\right)^{4x}} = 33$$

> Apply $(A^M)^N = A^{M \cdot N}$

$$\frac{2^{2x} \cdot 2^{4 \cdot 3x}}{2^{3 \cdot 4x}} = 33$$

$$\frac{2^{2x} \cdot 2^{12x}}{2^{12x}} = 33$$

> Apply $A^M \cdot A^N = A^{M+N}$

From this point we could have cancelled both of the 2^{12x} at the numerator and denominator:

$\frac{2^{2x} \cdot 2^{12x}}{2^{12x}} = 33$

$2^{2x} = 33$

And then proceed at the step pointed with the bellow arrow

$2^{14x}/2^{12x} = \frac{2^{14x}}{2^{12x}}$

We were just running out of space

$$\frac{2^{2x+12x}}{2^{12x}} = 33$$

$$2^{14x}/2^{12x} = 33$$

> Apply $\frac{A^M}{A^N} = A^{M-N}$

$$2^{14x-12x} = 33$$

$\log_2(33)/2 = \frac{\log_2(33)}{2}$

Again, we were getting to the end of the page

$$2^{2x} = 33$$

> Moves as a logarithm

$$2x = \log_2(33)$$

> Moves as division

$$x = \log_2(33)/2 \quad \text{»»} \quad x = \log_2(33)/2 \quad \text{»» Answer}$$

Number in red is the operation performed in the current step
Number in blue is the operation performed in the previous step

Problem 96) $\dfrac{3^{2x} \cdot 6^{2x}}{2^{2x}} = 45$

$$\frac{3^{2x} \cdot 6^{2x}}{2^{2x}} = 45$$

$$\frac{3^{2x} \cdot 6^{2x}}{2^{2x}} = 45$$

Apply $A^M \cdot B^M = (A \cdot B)^M$

As long as the exponents are EXACTLY the same, this rule applies

$$\frac{(3 \cdot 6)^{2x}}{2^{2x}} = 45$$

$$\frac{18^{2x}}{2^{2x}} = 45$$

Apply $\dfrac{A^M}{B^M} = \left(\dfrac{A}{B}\right)^M$

Yes, the rule also works for divisions as long as the exponents are EXACTLY the same

$$\left(\frac{18}{2}\right)^{2x} = 45$$

$$9^{2x} = 45$$

Moves as a logarithm

$$2x = \log_9(45)$$

Moves as division

$$x = \frac{\log_9(45)}{2}$$

$$x = \frac{\log_9(45)}{2} \quad \text{»» } \textbf{\textit{Answer}}$$

Number in red is the operation performed in the current step
Number in blue is the operation performed in the previous step

Problem 97) $\frac{5^{3x}}{25^{2x}} = 50$

$$\frac{5^{3x}}{25^{2x}} = 50$$

$$\frac{5^{3x}}{25^{2x}} = 50$$

Force all elements with x in the exponent to have the same base

We are still here:

$25 = 5 \cdot 5 = 5^2$

$$\frac{5^{3x}}{(5^2)^{2x}} = 50$$

$$\frac{5^{3x}}{(5^2)^{2x}} = 50$$

Apply $(A^M)^N = A^{M \cdot N}$

$$\frac{5^{3x}}{5^{2 \cdot 2x}} = 50$$

$$\frac{5^{3x}}{5^{4x}} = 50$$

Apply $\frac{A^M}{A^N} = A^{M-N}$

$5^{3x-4x} = 50$

$5^{-x} = 50$

A reminder about add/subtract rules. Different signs mean to perform a subtraction and keep the sign of the larger number. In this case, the -4x

$3x - 4x = -\,-1x = -x$

Moves as a logarithm

$-x = \log_5(50)$

Move the negative to the other side

$x = -\log_5(50)$

$x = -\log_5(50)$ »» **Answer**

Problem 98) $\frac{1}{7^{3x}} = 27$

$$\frac{1}{7^{3x}} = 27$$

$$\frac{1}{7^{3x}} = 27$$

$$\boxed{Apply \ \frac{1}{A^M} = A^{-M}}$$

$$7^{-3x} = 27$$

> We'll say it as many times as it is needed; x's in the denominator are a problem!

$$7^{-3x} = 27$$

$$\boxed{\text{Moves as a logarithm}}$$

$$-3x = \log_7(27)$$

$$\boxed{\text{Moves as division}}$$

$$x = \frac{\log_7(27)}{-3}$$

$$x = -\frac{\log_7(27)}{3} \quad \text{»» } \textbf{\textit{Answer}}$$

Problem 99) $3^{4x} = -8$

$$3^{4x} = -8$$

$$3^{4x} = -8$$

$$\boxed{\text{Moves as a logarithm}}$$

$$4x = \log_3(-8)$$

Logarithms of negative numbers are not defined by the rules of math (or at least not yet), therefore this problem can't be solved since we can't take the logarithm of -8

No Answer

Logarithm method: practice problems

1) $4^{2x-1} = 12$ 2) $3^{x+1} = 8$ 3) $\frac{1}{5^{3x+8}} = 15$

4) $6^{\frac{2x^2}{3}} = 35$ 5) $\frac{7^{8x} \cdot 49^{3x}}{343^{4x}} = 56$ 6) $\frac{2^{3x} \cdot 8^x}{64^x} = 28$

Number in red is the operation performed in the current step
Number in blue is the operation performed in the previous step

Logarithms method: dealing with negatives

As we saw on problem 99, the logarithm of a negative number does not exist. Anytime you are solving a problem and end up with a logarithm of a negative number either the problem has no solution or you did something wrong somewhere. But what happens when the negative is linked to the element having the x as an exponent? like this:

$$(-2)^x = 16$$

This problem requires a little bit of thinking. Negative numbers turn to positive when the exponent is even and remain negative when the exponent is odd:

$$(-2)^4 = 16 \rightarrow \text{Even exponent, positive answer}$$

But

$$(-2)^3 = -8 \rightarrow \text{Odd exponent, positive answer}$$

On the other side, positive numbers remain positive regardless their exponent:

$$2^4 = 16 \rightarrow \text{Even exponent, positive answer}$$

And

$$2^3 = 8 \rightarrow \text{Odd exponent, positive answer}$$

So we would have to look at whether the exponent is even or odd, the answer is positive or negative, in order to check whether the problem has a solution or not. Sounds like too much troubles.

To make the feasibility check easier, we developed a rule. You will need a calculator with the "log" button:

Pluto coefficient

$$For: (-A)^M = B \rightarrow Apply \rightarrow P = \left(\frac{log(|B|)}{log(|A|)}\right) \cdot (sign\ of\ B)$$

P has to be a whole number, if P has decimal: the problem has no solution.

If P is equal to a positive even number or a negative odd number: The problem has a solution

If P is equal to a negative even or a positive odd: The problem has no solution

Number in red is the operation performed in the current step
Number in blue is the operation performed in the previous step

142

Problem 100) $(-4)^{2x-3} = 256$

$$(-4)^{2x-3} = 256$$

$$(-4)^{2x-3} = +256$$

$$|A| = |-4| = 4 \; ; |B| = |256| = 256 \; ; sign\ of\ B = +$$

The base having the exponent of x is a negative number. We have to use the Pluto coefficient to check if the problem has a solution

Identify the elements of the Pluto coefficient

$$P = \frac{\log(|B|)}{\log(|A|)} \cdot (sign\ of\ B) \to P = \frac{\log(256)}{\log(4)} \cdot (+)$$

Plug the values into the formula

$$P = \frac{2.408}{0.602} \cdot (+) \to P = +4$$

P is a positive even number. The problem can be solved!

$$4^{2x-3} = 256$$

$$4^{2x-3} = 256$$

Once the P coefficient has proven the problem can be solved, we can (and it is recommended) treat the base as a positive number

Moves as a logarithm

$$2x - 3 = \log_4(256)$$

$$2x - 3 = 4$$

Moves as positive

$$2x = 4 + 3$$

Moves as division

$$x = \frac{7}{2}$$

Yes, we could have solved this problem using the similar bases method because: $256 = 4 \cdot 4 \cdot 4 \cdot 4 = 4^4$ Cancel the bases and continue at this step

Actually $\log_4(256) = 4$ since it is a number without decimals we just leave it as 4

$$x = \frac{7}{2} \; »» \; \textbf{\textit{Answer}}$$

Number in red is the operation performed in the current step
Number in blue is the operation performed in the previous step

Problem 101) $(-8)^{3x+5} = 512$

$$(-8)^{3x+5} = 512$$

$$(-8)^{3x+5} = +512$$

$$|A| = |-8| = 8 \; ; |B| = |512| = 512; sign\ of\ B = +$$

Every time the base of the x is a negative number we must check if the problem can be solved

Identify the elements of the Pluto coefficient

$$P = \frac{\log(|B|)}{\log(|A|)} \cdot (sign\ of\ B) \to P = \frac{\log(512)}{\log(8)} \cdot (+)$$

Plug the values into the formula

$$P = \frac{2.709}{0.903} \cdot (+) \to P = +3$$

P is a positive odd number. The problem can NOT be solved!

No Answer

Problem 102) $(-10)^{5x} = 250$

$$(-10)^{5x} = 250$$

$$(-10)^{5x} = +250$$

$$|A| = |-10| = 10 \; ; |B| = |250| = 250 \; ; sign\ of\ B = +$$

Again, a negative base for the x in the exponent, we must check if the problem has solution

Identify the elements of the Pluto coefficient

$$P = \frac{\log(|B|)}{\log(|A|)} \cdot (sign\ of\ B) \to P = \frac{\log(250)}{\log(10)} \cdot (+)$$

Plug the values into the formula

$$P = \frac{2.398}{1} \cdot (+) \to P = +2.398$$

P has decimals, it doesn't matter the sign.

The problem can NOT be solved!

No Answer

Yeah when we use the log button in the calculator is like using $\log_{10}()$
Therefore
$\log(10) = \log_{10}(10) = 1$

Number in red is the operation performed in the current step
Number in blue is the operation performed in the previous step

Problem 103) $(-3)^{4x+2} = -27$

$$(-3)^{4x+2} = -27$$

$$(-3)^{4x+2} = -27$$

$$|A| = |-3| = 3 \; ; |B| = -27 = 27 \; ; sign \; of \; B = -$$

Negative base = check if the problem has solution

Identify the elements of the Pluto coefficient

$$P = \frac{\log(|B|)}{\log(|A|)} \cdot (sign \; of \; B) \rightarrow P = \frac{\log(27)}{\log(3)} \cdot (-)$$

Plug the values into the formula

$$P = \frac{1.431}{0.477} \cdot (-) \rightarrow P = -3$$

P is a negative odd number.

The problem can be solved!

Once the P coefficient has proven the problem can be solved, we can (and it is recommended) treat the base as a positive number

$$3^{4x+2} = 27$$

$$3^{4x+2} = 27$$

Moves as a logarithm

$$4x + 2 = \log_3(27)$$

Again, the similar bases method can be used, just look:

$$27 = 3 \cdot 3 \cdot 3 = 3^3$$

Cancel the bases and continue at this step

$$4x + 2 = 3$$

Moves as negative

$$4x = 3 - 2$$

Actually $\log_3(27) = 3$ since it is a number without decimals we just leave it as 3

Moves as division

$$x = \frac{1}{4}$$

$$x = \frac{1}{4} \;\; \text{»» } \textbf{\textit{Answer}}$$

Problem 104) $(-5)^{4x-1} = -625$

$$(-5)^{4x-1} = -625$$

$$(-5)^{4x-1} = -625$$

$$|A| = |-5| = 5 \; ; |B| = |-625| = 625 \; ; sign \; of \; B = -$$

We won't even say it this time. You know what you have to do with the negative base of x

Identify the elements of the Pluto coefficient

$$P = \frac{\log(|B|)}{\log(|A|)} \cdot (sign \; of \; B) \rightarrow P = \frac{\log(625)}{\log(5)} \cdot (-)$$

Plug the values into the formula

$$P = \frac{2.796}{0.699} \cdot (-) \rightarrow P = -4$$

P is a negative even. The problem can NOT be solved!

No Answer

Problem 105) $(2)^{7x} = -32$

$$(2)^{7x} = -32$$

$$(2)^{7x} = -32$$

The exponent of x has a positive base, therefore the Pluto coefficient is not necessary

We have a similar case as problem 99. We can solve it in the same manner and realize it has no solution or we could use the following information:

Positive numbers can't be turned into negative numbers by using exponents.

It doesn't matter the value of x, POSITIVE 2 will never turn to NEGATIVE 32, therefore the problem has no solution

No Answer

Dealing with negatives: practice problems

1) $(-10)^x = 1,000,000$ 2) $(-7)^x = -343$

3) $(-3)^{2x} = 243$ 4) $(-9)^{4x-1} = 6561$

5) $(-4)^{x+5} = -1024$ 6) $(-6)^{6-x} = -36$

Number in red is the operation performed in the current step
Number in blue is the operation performed in the previous step

We've learned that logarithms are the opposite of exponents and we used that information to solve for problems involving x in the exponent position. It is logical to think that an exponent will be used to solve for logarithms, you would be correct. The reason we have created a whole new chapter about logarithms is because they have their own rules and their behavior requires a bit more thinking than exponents even though they are just the opposite. The main rule will be, for any x inside of the logarithm we can move the base of the logarithm as the base of an exponent on the other side of the equation in order to solve for the equation. Of course we have an example, math is easier with visual explanations:

Example 1

$$\log_2(x) = 4$$

Moves as the base of an exponent

$$x = 2^4$$

As usual, when we are facing problems involving different elements, we will have to group all of the x into only one before we can solve for the logarithm. In order to be able to group x's within logarithms we must learn the rules of logarithms first:

Addition of logarithms = Multiply the inside of both logarithms and keep only one

$$\log(A) + \log(B) = \log(A \cdot B) \rightarrow Example \rightarrow \log(3) + \log(4) = \log(3 \cdot 4) = \log(12)$$

Subtraction of logarithms = Divide the inside of both logarithms and keep only one

$$\log(A) - \log(B) = \log\left(\frac{A}{B}\right) \rightarrow Example \rightarrow \log(8) - \log(2) = \log\left(\frac{8}{2}\right) = \log(4)$$

Important note: the value inside of the second logarithm MUST be the denominator of the fraction

For the two rules above the logarithms must have the same base, otherwise it does not work

Exponent inside of the logarithm = bring down the exponent as a multiplication

$$\log(A^M) = M \cdot \log(A) \rightarrow Example \rightarrow \log(4^5) = 5 \cdot \log(4)$$

Logarithm of 1 = equals 0 every time, regardless of the base

$$\log_A(1) = 0 \rightarrow Example \rightarrow \log_{15}(1) = 0$$

The logarithm of negative numbers or the logarithm of zero does not exist, meaning there is no answer.

Some of these logarithm problems have different methods to get the final answer. We will show you a couple every time there is a chance.

Number in red is the operation performed in the current step
Number in blue is the operation performed in the previous step

There is one more concept we need to learn before we can get into logarithms; Euler's number. You've probably heard about it before, it is identified by the letter e. Its actual value is 2.718... with an infinite amount of decimals. Same as the famous pi we'll, never find two equal decimals next to each other. It is one of the most important irrational numbers and it is commonly used in areas like physics, chemistry and mathematics itself to calculate and demonstrate many different phenomena. For the problems in this book you only need to remember it is represented by the letter e, it is a constant number and it is the opposite of the natural logarithm.

In your calculator (and in many math problems) you might see something such as ln which is the natural logarithm. Let us show you what the ln represents:

$$\ln(A) \rightarrow \log_e(A)$$

It is literally just that; ln is the same as having a logarithm with base e

So if we ever encounter a problem in which we have to solve for x inside of ln like this one:

Example 2:

$$\ln(x) = 3$$

We first convert ln into a log with base e
$$\log_e(x) = 3$$

And then apply the opposite rule
$$\log_e(x) = 5$$

Moves as the base of an exponent

$$x = e^5$$

Number in red is the operation performed in the current step
Number in blue is the operation performed in the previous step

148

Problem 106) $\log_3(x) = 4$

$$\log_3(x) = 4$$

$$\log_3(x) = 4$$

Moves as the base of an exponent

$$x = 3^4$$

$$x = \boldsymbol{81} \text{ »» } \boldsymbol{Answer}$$

Problem 107) $\log_5(x) = 3$

$$\log_5(x) = 3$$

$$\log_5(x) = 3$$

Moves as the base of an exponent

$$x = 5^3$$

$$x = \boldsymbol{125} \text{ »» } \boldsymbol{Answer}$$

Problem 108) $\ln(x) = 2$

$$\ln(x) = 2$$

$$\ln(x) = 2$$

Apply $\ln(x) = \log_e(x)$

$$\log_e(x) = 2$$

$$\log_e(x) = 2$$

Moves as the base of an exponent

$$x = e^2$$

$$x = \boldsymbol{e^2} \text{ »» } \boldsymbol{Answer}$$

Same as roots, this will be a very long answer. We can leave it like that

Number in red is the operation performed in the current step
Number in blue is the operation performed in the previous step

Problem 109) $\ln(x) - 2 = 5$

$$\ln(x) - 2 = 5$$

$$\ln(x) - 2 = 5$$

Same as all other chapters, we must first isolate the term containing the x

Moves as positive

$$\ln(x) = 5 + 2$$

Apply $\ln(x) = \log_e(x)$

$$\log_e(x) = 7$$

$$\log_e(x) = 7$$

Same as roots, this will be a very long answer. We can leave it like that

Moves as the base of an exponent

$$x = e^7$$

$$x = e^7 \text{ »» } \textbf{\textit{Answer}}$$

Problem 110) $3\ln(x) = 12$

$$3\ln(x) = 12$$

$$3\ln(x) = 12$$

Moves as division

Always isolate the term having the x. It is safe to play Like this

$$\ln(x) = \frac{12}{3}$$

Apply $\ln(x) = \log_e(x)$

$$\log_e(x) = 4$$

$$\log_e(x) = 4$$

Moves as the base of an exponent

$$x = e^4$$

$$x = e^4 \text{ »» } \textbf{\textit{Answer}}$$

Number in red is the operation performed in the current step
Number in blue is the operation performed in the previous step

Problem 111) $\ln\left(\frac{1}{x}\right) + 5 = 3$

$$\ln\left(\frac{1}{x}\right) + 5 = 3$$

$$\ln\left(\frac{1}{x}\right) + 5 = 3$$

Moves as negative

Method 1

$$\ln\left(\frac{1}{x}\right) = 3 - 5$$

Apply $\ln(x) = \log_e(x)$

$$\log_e\left(\frac{1}{x}\right) = -2$$

$$\log_e\left(\frac{1}{x}\right) = -2$$

Moves as the base of an exponent

$$\frac{1}{x} = e^{-2}$$

$$\frac{1}{x} = e^{-2}$$

Apply $A^{-M} = \frac{1}{A^M}$

$$\frac{1}{x} = \frac{1}{e^2}$$

Don't ever forget, x's in the denominator are problematic

Flip both sides of the equation

$$x = e^2$$

$x = e^2$ »» **Answer**

Yes, this is allowed as long as we do it on both sides of the equation at the same time

Number in red is the operation performed in the current step
Number in blue is the operation performed in the previous step

Method 2

$$\ln\left(\frac{1}{x}\right) = -2$$

Apply $\dfrac{1}{A^M} = A^{-M}$

$$\ln(x^{-1}) = -2$$

Apply $\ln(x) = \log_e(x)$

$$\log_e\left(x^{-1}\right) = -2$$

Apply $\log(A^M) = M \cdot \log(A)$

$$-1 \cdot \log_e(x) = -2$$

$$-1 \cdot \log_e(x) = -2$$

Moves as division

$$\log_e(x) = \frac{-2}{-1}$$

Moves as the base of an exponent

Recap from earlier chapters: Two negatives in the fraction are equal to a positive answer

$$x = e^2$$

$$x = e^2 \text{ »» } \textbf{\textit{Answer}}$$

Number in red is the operation performed in the current step

Number in blue is the operation performed in the previous step

Problem 112) $3\log_2(x^4) = 48$

$$3\log_2(x^4) = 48$$

Method 1

$$3\log_2(x^4) = 48$$

Moves as division

$$\log_2(x^4) = \frac{48}{3}$$

Moves as the base of an exponent

$$x^4 = 2^{16}$$

Moves as a root

$$x = \sqrt[4]{65536}$$

$$x = 16 \text{ »» } \textbf{\textit{Answer}}$$

Method 2

$$3\log_2(x^4) = 48$$

Apply $\log(A^M) = M \cdot \log(A)$

$$4 \cdot 3\log_2(x)$$

$$12\log_2(x) = 48$$

Moves as division

$$\log_2(x) = {48}/{12}$$

$$x = 2^4$$

$$x = 16 \text{ »» } \textbf{\textit{Answer}}$$

Regardless the method selected, our main goal is to isolate the element having the x

Number in red is the operation performed in the current step
Number in blue is the operation performed in the previous step

Problem 113) $6\log_4(\sqrt{x}) = 9$

$$6\log_4(\sqrt{x}) = 9$$

Method 1

$$6\log_4(\sqrt{x}) = 9$$

Moves as division

One more time, simplify your fractions:

$$\frac{9 \div 2}{6 \div 2} = \frac{3}{2}$$

$$\log_4(\sqrt{x}) = \frac{9}{6}$$

Moves as the base of an exponent

$$\sqrt[2]{x} = 4^{\frac{3}{2}}$$

Do not ever forget that:

$$\sqrt{x} = \sqrt[2]{x}$$

$$\sqrt[2]{x} = 8$$

Moves as an exponent

$$x = 8^2$$

$$x = 64 \ \text{»» } Answer$$

In order to turn this into a number let's remember this rule:

$$A^{\frac{M}{N}} = \sqrt[N]{A^M}$$

Apply it to the problem

$$4^{\frac{3}{2}} = \sqrt[2]{4^3} = \sqrt[2]{64} = 8$$

Number in red is the operation performed in the current step
Number in blue is the operation performed in the previous step

154

Method 2

$$6 \log_4\left(\sqrt{x}\right) = 9$$

$$\boxed{Apply \; A^{\frac{M}{N}} = \sqrt[N]{A^M}}$$

$$6 \log_4\left(x^{\frac{1}{2}}\right) = 9$$

$$6 \log_4\left(x^{\frac{1}{2}}\right) = 9$$

$$\boxed{Apply \; \log(A^M) = M \cdot \log(A)}$$

Just in case. Whenever a root does not have a number, assume a square root. Whenever a number has no exponent, assume an exponent of 1. Therefore:

$$\sqrt{x} = \sqrt[2]{x^1} = x^{\frac{1}{2}}$$

$$\frac{1}{2} \cdot 6 \log_4(x) = 9$$

$$\frac{6}{2} \log_4(x) = 9$$

$$3 \log_4(x) = 9$$

$$\boxed{Moves \; as \; division}$$

$$\log_4(x) = \frac{9}{3}$$

$$\boxed{Moves \; as \; the \; base \; of \; an \; exponent}$$

$$x = 4^3$$

$$x = 64 \; »» \; \textbf{Answer}$$

Number in red is the operation performed in the current step
Number in blue is the operation performed in the previous step

155

Problem 114) $4 \log_5(3x^2 - 1) + 10 = 22$

$$4 \log_5(3x^2 - 1) + 10 = 22$$

$$4 \log_5(3x^2 - 1) + 10 = 22$$

And we are still using the "furthest away first" rule

There is a complex equation inside of the logarithm, but then again, the x is inside of the logarithm, let's isolate it

Moves as negative

$$4 \log_5(3x^2 - 1) = 22 - 10$$

Moves as division

$$\log_5(3x^2 - 1) = \frac{12}{4}$$

Moves as the base of an exponent

$$3x^2 - 1 = 5^3$$

Moves as positive

$$3x^2 = 125 + 1$$

Moves as division

$$x^2 = \frac{126}{3}$$

Moves as a root

$$x = \pm \sqrt[2]{42}$$

$$x = \sqrt{42} \text{ and } x = -\sqrt{42} \text{ »» } \textbf{Answer}$$

Friendly reminder about square roots having both positive and negative answers

Number in red is the operation performed in the current step
Number in blue is the operation performed in the previous step

156

Problem 115) $3 - 7\ln(4x - 2) = 10$

$$3 - 7\ln(4x - 2) = 10$$

$$3 - 7\ln(4x - 2) = 10$$

Moves as negative

As long as that x is inside of the logarithm the problem can't be solved, so first, isolate the logarithm

$$-7\ln(4x - 2) = 10 - 3$$

Moves as division

$$\ln(4x - 2) = \frac{7}{-7}$$

$$Apply \ \ln(x) = \log_e(x)$$

$$\log_e(4x - 2) = -1$$

Now that the logarithm is by itself, we can remove it to get the x outside

$$\log_e(4x - 2) = -1$$

Moves as the base of an exponent

$$4x - 2 = e^{-1}$$

$$4x - 2 = e^{-1}$$

$$Apply \ \frac{1}{A^M} = A^{-M}$$

$$4x - 2 = \frac{1}{e^1}$$

Moves as positive

It is not mandatory to write exponents of 1, so:

$$\frac{1}{e^1} = \frac{1}{e}$$

$$4x = \frac{1}{e} + 2$$

For addition of fractions apply:
$$\frac{A}{B} + \frac{C}{D} = \frac{A \cdot D + B \cdot C}{B \cdot D}$$
$$\frac{1}{e} + \frac{2}{1} = \frac{1 \cdot 1 + e \cdot 2}{e \cdot 1}$$
$$= \frac{1 + 2e}{e}$$

Moves as division

$$x = \frac{1 + 2e}{e \cdot 4}$$

$$x = \frac{1 + 2e}{4e} \ \text{»» } \boldsymbol{Answer}$$

Problem 116) $2\ln(1 + 5x) + 12 = 20$

$$2\ln(1 + 5x) + 12 = 20$$

$$2\ln(1 + 5x) + 12 = 20$$

> Moves as negative

$$2\ln(1 + 5x) = 20 - 12$$

> Moves as division

$$\ln(1 + 5x) = \frac{8}{2}$$

> *Apply* $\ln(x) = \log_e(x)$

$$\log_e(1 + 5x) = 4$$

$$\log_e(1 + 5x) = 4$$

> Moves as the base of an exponent

$$1 + 5x = e^4$$

> Moves as negative

$$5x = e^4 - 1$$

> Moves as division

$$x = \frac{e^4 - 1}{5}$$

$$x = \frac{e^4 - 1}{5} \quad \text{»» } \textbf{\textit{Answer}}$$

Problem 117) $\log_3(3x^2) + \log_3(5x) = 2$

$$\log_3(3x^2) + \log_3(5x) = 2$$

$$log_3(3x^2) + log_3(5x) = 2$$

Apply $\log(A) + \log(B) = \log(A \cdot B)$

This rule is only possible because the two logarithms have the SAME base (in this case base 3)

There are too many logarithms containing x, we have to first group Them all together

$$\log_3(3x^2 \cdot 5x) = 2$$

$$\log_3(15x^3) = 2$$

Multiply x with x and number with number:

$3x^2 \cdot 5x$
$= (3 \cdot 5)(x^2 \cdot x)$
$= 15x^3$

Moves as the base of an exponent

$$15x^3 = 3^2$$

Moves as division

$$x^3 = \frac{9}{15}$$

One more time, simplify your fractions:

$$\frac{9 \div 3}{15 \div 3} = \frac{3}{5}$$

Moves as a root

$$x = \sqrt[3]{\frac{3}{5}}$$

$$x = \sqrt[3]{\frac{3}{5}} \text{ »» } \textbf{\textit{Answer}}$$

Just another friendly reminder about leaving answers with the root and fractions if the actual answer has too many decimals like this one

Number in red is the operation performed in the current step
Number in blue is the operation performed in the previous step

159

Problem 118) $3\log_4(2x) + 5\log_4(x^2) = 4$

$$3\log_4(2x) + 5\log_4(x^2) = 4$$

$$3\log_4(2x) + 5\log_4(x^2) = 4$$

> Yes, we have to group the logarithms but we can't really do that with that 3 and that 5 multiplying outside. We need to bring them inside the logarithm first

$$\boxed{\textit{Apply } \log(A^M) = M \cdot \log(A)}$$

$$\log_4\left((2x)^3\right) + \log_4\left((x^2)^5\right) = 4$$

$$\log_4\left((2x)^3\right) + \log_4\left((x^2)^5\right) = 4$$

$$\boxed{\textit{Apply } (A^M)^N = A^{M \cdot N}}$$

$$\log_4(2^{1 \cdot 3} x^{1 \cdot 3}) + \log_4(x^{2 \cdot 5}) = 4$$

$$\log_4(8x^3) + \log_4(x^{10}) = 4$$

$$\boxed{\textit{Apply } \log(A) + \log(B) = \log(A \cdot B)}$$

> Multiply x with x and number with number:
>
> $$8x^3 \cdot x^8$$
> $$= (8 \cdot 1)(x^3 \cdot x^{10})$$
> $$= 8x^{13}$$

$$\log_4(8x^3 \cdot x^{10}) = 4$$

$$\log_4(8x^{13}) = 4$$

$$\boxed{\text{Moves as the base of an exponent}}$$

$$8x^{13} = 4^4$$

$$\boxed{\text{Moves as division}}$$

$$x^{13} = \frac{256}{8}$$

$$\boxed{\text{Moves as a root}}$$

$$x = \sqrt[13]{32}$$

$$x = \sqrt[13]{32} \text{ »» } \textbf{\textit{Answer}}$$

Number in red is the operation performed in the current step

Number in blue is the operation performed in the previous step

Problem 119) $\ln(3x) + 20 = 23 - 2\ln(5x^2) + \ln(15x^3)$

$$\ln(3x) + 20 = 23 - 2\ln(5x^2) + \ln(15x^3)$$

Moves as negative

Let's start by putting all logarithms in the same side of the equation

$$\ln(3x) + 20 = 23 - 2\,ln\!\left(5x^2\right) + ln\!\left(15x^3\right)$$

Moves as positive

$$\ln(3x) + 2\ln(5x^2) - \ln(15x^3) + 20 = 23$$

Moves as negative

We are still not able to group logarithms, that 2 in red outside of one of them is disturbing us

And now move all none logarithm/none x elements to the other side

$$\ln(3x) + 2\ln(5x^2) - \ln(15x^3) = 23 - 20$$

Apply $\log(A^M) = M \cdot \log(A)$

$$\ln(3x) + \ln\!\left((5x^2)^2\right) - \ln(15x^3) = 3$$

$$\ln(3x) + \ln\!\left((5x^2)^2\right) - \ln(15x^3) = 3$$

Apply $(A^M)^N = A^{M \cdot N}$

$$\ln(3x) + \ln(5^{1\cdot2}x^{2\cdot2}) - \ln(15x^3) = 3$$

Finally, we can group all logarithms into only one

$$\ln(3x) + \ln(25x^4) - \ln(15x^3) = 3$$

Apply $\log(A) + \log(B) = \log(A \cdot B)$

$$\ln(3x \cdot 25x^4) - \ln(15x^3) = 3$$

$$\ln(75x^4) - \ln(15x^3) = 3$$

Apply $\log(A) - \log(B) = \log\!\left(\dfrac{A}{B}\right)$

$$\ln\!\left(\frac{75x^5}{15x^3}\right) = 3$$

Number in red is the operation performed in the current step
Number in blue is the operation performed in the previous step

161

$$\ln\left(\frac{75x^5}{15x^3}\right) = 3$$

$$\ln\left(\frac{75}{15} \cdot \frac{x^5}{x^3}\right) = 3$$

Same case as the multiplication, we can divide the numbers with the numbers and a separate division for the x's

$$\ln(5x^2) = 3$$

Apply $\ln(x) = \log_e(x)$

$$\log_e(5x^2) = 3$$

$$\log_e(5x^2) = 3$$

Moves as the base of an exponent

$$5x^2 = e^3$$

Moves as division

$$x^2 = \frac{e^3}{5}$$

Moves as a root

Don't ever forget, roots have both positive and negative answers

$$x = \pm\sqrt[2]{\frac{e^3}{5}}$$

$$x = \sqrt{\frac{e^3}{5}} \ \textbf{and} \ x = -\sqrt{\frac{e^3}{5}} \ \text{»» } \textbf{Answer}$$

Number in red is the operation performed in the current step
Number in blue is the operation performed in the previous step

Problem 120) $5 - 6\log_3(2x^4) = \log_3(5x^4) - 2\log_3(8x^{12}) + 1$

$$5 - 6\log_3(2x^4) = \log_3(5x^4) - 2\log_3(8x^{12}) + 1$$

$$5 - 6\log_3(2x^4) = \log_3(5x^4) - 2\log_3(8x^{12}) + 1$$

First, group all logarithms on the same side... yes we can group them on the right side!

Moves as positive

$$5 = \log_3(5x^4) + 6\log_3(2x^4) - 2\log_3(8x^{12}) + 1$$

Moves as negative

As a general rule, we can't put logarithms together if they have number multiplying them

$$5 - 1 = \log_3(5x^4) + 6\log_3(2x^4) - 2\log_3(8x^{12})$$

Apply $\log(A^M) = M \cdot \log(A)$

$$4 = \log_3(5x^4) + \log_3\left((2x^4)^6\right) - \log_3\left((8x^{12})^2\right)$$

$$4 = \log_3(5x^4) + \log_3\left((2x^4)^6\right) - \log_3\left((8x^{12})^2\right)$$

Apply $(A^M)^N = A^{M \cdot N}$

$$4 = \log_3(5x^4) + \log_3(2^{1 \cdot 6}x^{4 \cdot 6}) - \log_3(8^{1 \cdot 2}x^{12 \cdot 2})$$

$$4 = \log_3(5x^4) + \log_3(64x^{24}) - \log_3(64x^{24})$$

There are two elements equal to each other but one is negative and one is positive, they can be cancelled

$$4 = \log_3(5x^4)$$

Moves as the base of an exponent

$$3^4 = 5x^4$$

Moves as division

$$81/_5 = x^4$$

Apply:

$$\sqrt{\frac{A}{B}} = \frac{\sqrt{A}}{\sqrt{B}}$$

$$\sqrt[4]{\frac{81}{5}} = \frac{\sqrt[4]{81}}{\sqrt[4]{5}} = \frac{3}{\sqrt[4]{5}}$$

Moves as a root

$$\sqrt[4]{81/_5} = x$$

$$x = {}^3/_{\sqrt[4]{5}} \quad \text{»» } \textbf{Answer}$$

Number in red is the operation performed in the current step
Number in blue is the operation performed in the previous step

Problem 121) $\log(6x^5) + \log(8x^6) - \log(5x^3) - \log(12x^5) = 2$

$$\log(6x^5) + \log(8x^6) - \log(5x^3) - \log(12x^5) = 2$$

$$\log(6x^5) + \log(8x^6) - \log(5x^3) - \log(12x^5) = 2$$

$$\log(6x^5) + \log(8x^6) - (\log(5x^3) + \log(12x^5)) = 2$$

$$\log(6x^5) + \log(8x^6) - (\log(5x^3) + \log(12x^5)) = 2$$

Apply $\log(A) + \log(B) = \log(A \cdot B)$

Having two negatives in a row might be confusing. Apply this rule:
$-A - B$
$= -(A + B)$

$$\log(6x^5 \cdot 8x^6) - \log(5x^3 \cdot 12x^5) = 2$$

Just in case you forgot, this is how we solved the left most logarithm:
$6x^5 \cdot 8x^6$
$= (6 \cdot 8)(x^5 \cdot x^6)$
$= 48 \cdot x^{5+6} = 48x^{11}$

$$\log(48x^{11}) - \log(60x^8) = 2$$

Apply $\log(A) - \log(B) = \log\left(\dfrac{A}{B}\right)$

Just in case you forgot, this is how we solved the right most logarithm:
$5x^3 \cdot 12x^5$
$= (5 \cdot 12)(x^3 \cdot x^5)$
$= 60 \cdot x^{3+5} = 60x^8$

$$\log\left(\frac{48x^{11}}{60x^8}\right) = 2$$

One more time:
$\dfrac{48x^{11}}{60x^8} = \dfrac{48}{60} \cdot \dfrac{x^{11}}{x^8}$
$= \dfrac{4}{5} \cdot x^{11-8} = \dfrac{4x^3}{5}$

$$\log\left(4x^3/5\right) = 2$$

$$\log_{10}\left(4x^3/5\right) = 2$$

Whenever a logarithm does not have a written base, it is assumed it has a base of 10
$\log(A) = \log_{10}(A)$

Moves as the base of an exponent

$$4x^3/5 = 10^2$$

Moves as multiplication

$$4x^3 = 100 \cdot 5$$

Moves as division

$$x^3 = 500/4$$

Moves as a root

$$x = \sqrt[3]{125}$$

$$x = 5 \text{ »» } \textbf{Answer}$$

Problem 122) $3\log(4x) - \log(8x^5) = -2$

$$3\log(4x) - \log(8x^5) = -2$$

$$3\log(4x) - \log(8x^5) = -2$$

| Apply $\log(A^M) = M \cdot \log(A)$ |

$$\log((4x)^3) - \log(8x^5) = -2$$

$$\log\left((4x)^3\right) - \log(8x^5) = -2$$

| Apply $A^M \cdot B^M = (A \cdot B)^M$ |

Remember, numbers multiplying outside of the logarithms are considered obstacles

$$\log(4^3 \cdot x^3) - \log(8x^5) = -2$$

$$\log(64x^3) - \log(8x^5) = -2$$

| Apply $\log(A) - \log(B) = \log\left(\dfrac{A}{B}\right)$ |

$$\log\left(\frac{64x^3}{8x^5}\right) = -2$$

$$\log\left(\frac{8}{x^2}\right) = -2$$

| Apply $\log(x) = \log_{10}(x)$ |

Again, we'll help you with the inside of the logarithm:
$$\frac{64x^3}{8x^5} = \frac{64}{8} \cdot \frac{x^3}{x^5}$$
$$= 8 \cdot x^{3-5} = 8x^{-2} = \frac{8}{x^2}$$

$$\log_{10}\left(\frac{8}{x^2}\right) = -2$$

$$\log_{10}\left(\frac{8}{x^2}\right) = -2$$

| Moves as the base of an exponent |

$$\frac{8}{x^2} = 10^{-2}$$

Number in red is the operation performed in the current step
Number in blue is the operation performed in the previous step

165

$$\frac{8}{x^2} = 10^{-2}$$

$$\frac{8}{x^2} = \frac{1}{100}$$

Get rid of negative exponents as soon as possible. Remember:

$$A^{-M} = \frac{1}{A^M}$$

Therefore:

$$10^{-2} = \frac{1}{10^2} = \frac{1}{100}$$

Flip both sides of the equation

A few chapters ago, we mentioned the fact about x's in the denominator being a problem and yes, they are still a problem. Also yes, we can flip fractions in an equation, as long as we flip both sides of the equation simultaneously

$$\frac{x^2}{8} = 100$$

Moves as multiplication

$$x^2 = 100 \cdot 8$$

Moves as a root

$$x = \sqrt[2]{800}$$

$$x = \sqrt{800} \ \textbf{and} \ x = -\sqrt{800} \ \text{»» } \textbf{Answer}$$

Number in red is the operation performed in the current step
Number in blue is the operation performed in the previous step

Problem 123) $\log_{10}\left(\frac{1}{x^2}\right) = -2$

$$\log_{10}\left(\frac{1}{x^2}\right) = -2$$

$$\log_{10}\left(\frac{1}{x^2}\right) = -2$$

$$\boxed{Apply \ \frac{1}{A^M} = A^{-M}}$$

$$\log_{10}(x^{-2}) = -2$$

$$\log_{10}(x^{-2}) = -2$$

$$\boxed{Apply \ \log(A^M) = M \cdot \log(A)}$$

$$-2\log_{10}(x) = -2$$

$$-2\log_{10}(x) = -2$$

$$\boxed{\text{Moves as division}}$$

$$\log_{10}(x) = \frac{-2}{-2}$$

$$\boxed{\text{Moves as the base of an exponent}}$$

$$x = 10^1$$

$$x = 10 \ \text{»» } \textbf{\textit{Answer}}$$

Double negative in the
fraction, the answer
must be positive:

$$\frac{-2}{-2} = +1 = 1$$

Number in red is the operation performed in the current step
Number in blue is the operation performed in the previous step

Problem 124) $\log_{10}\left(\frac{1}{4x^2}\right) = -2$

$$\log_{10}\left(\frac{1}{4x^2}\right) = -2$$

$$\log_{10}\left(\frac{1}{4x^2}\right) = -2$$

The logarithm is already by itself, we can start solving the problem by moving it to the other side. There are like three paths to solve this problem. We'll show you this one, the other two, well, that would be a good practice. If you get the same final answer, then you are correct

Moves as the base of an exponent

$$\frac{1}{4x^2} = 10^{-2}$$

$$\frac{1}{4x^2} = 10^{-2}$$

Apply $\frac{1}{A^M} = A^{-M}$

$$\frac{1}{4x^2} = \frac{1}{10^2}$$

$$\frac{1}{4x^2} = \frac{1}{100}$$

Flip both sides of the equation

$$4x^2 = 100$$

Agh! An x in the denominator! Let's flip both sides of the equation

Moves as division

$$x^2 = \frac{100}{4}$$

Moves as a root

$$x = \sqrt[2]{25}$$

$$x = \pm 5$$

$$x = 5 \ and \ x = -5 \ \text{»» } \textbf{\textit{Answer}}$$

Number in red is the operation performed in the current step
Number in blue is the operation performed in the previous step

168

Problem 125) $\log(x^5) \cdot \log_5(7x^3) = 5$

$$\log(x^5) \cdot \log_5(7x^3) = 5$$

The bases of the logarithms are not equal. We can't put them together if their bases are not equal.

This problem can't be solved using algebra

No Answer

Problem 126) $\log_3(5x^2) + \log_4(3x^2) = 10$

$$\log_3(5x^2) + \log_4(3x^2) = 10$$

$$\log_3(5x^2) + \log_4(3x^2) = 10$$

The bases of the logarithms are not equal. We can't put them together if their bases are not equal.

This problem can't be solved using algebra

No Answer

Problem 127) $\log_6(3x^2) - \log_8(\sqrt{x}) = 20$

$$\log_6(3x^2) - \log_8(\sqrt{x}) = 20$$

$$\log_6(3x^2) - \log_8(\sqrt{x}) = 20$$

The bases of the logarithms are not equal. We can't put them together if their bases are not equal.

This problem can't be solved using algebra. Yes, we copy and pasted again.

No Answer

Number in red is the operation performed in the current step
Number in blue is the operation performed in the previous step

171

Problem 130) $\sqrt[3]{32 \log(x^2)} = 4$

$\sqrt[3]{32 \log(x^2)} = 4$

$\sqrt[3]{32 \log(x^2)} = 4$

| Moves as a root |

$32 \log(x^2) = 4^3$

> Same as the exponent, the root is outside of the logarithm. We must move the root first

| Moves as division |

$\log(x^2) = \dfrac{64}{32}$

| Apply $\log(A^M) = M \cdot \log(A)$ |

$2 \cdot \log(x) = 2$

$2 \cdot \log(x) = 2$

| Moves as division |

$\log(x) = \dfrac{2}{2}$

| Apply $\log(x) = \log_{10}(x)$ |

> And now that the logarithm is by itself, it can be moved

$\log_{10}(x) = 1$

| Moves as the base of an exponent |

$x = 10^1$

$x = 10$ »» **Answer**

Number in red is the operation performed in the current step
Number in blue is the operation performed in the previous step

172

X in the logarithm: practice problems

1) $\ln(x) = 1$ 2) $\log_3(x) = -1$

3) $\log_4(x) = -1$ 4) $\log_{10}(x) = 1$

5) $\log_{10}\left(\dfrac{1}{x}\right) = 2$ 6) $4\ln(x) - 3 = 5$

7) $\dfrac{3\log_2(x)}{2} + 1 = 4$ 8) $6\log_3(x + 2) - 1 = 5$

9) $-2\log(x^2) + 4 = 8$ 10) $3\ln(2\sqrt{x}) - 1 = 8$

11) $\ln(2x^4) + \ln(8x) = 2$ 12) $\log_4(15x^3) - \log_4(5x^2) = 3$

13) $2\log(4x^2) - 3\log(6x) = 1$

14) $5\ln(2x^4) - 4\ln(x^4) + 2\ln(6x^5) - 6\ln(2x^2) = 3$

Number in red is the operation performed in the current step
Number in blue is the operation performed in the previous step

173

Chapter six:
Special cases

So far we have learned enough to be considered fluent in the language of math, or at least for the algebra part. The techniques explained up to this point can help us solve the vast majority of the problems we can face in algebra. Now we are going to learn some unique scenarios, situations that could probably be solved using the methods learned in the first 130 problems of this book but they have other alternatives. These alternative solutions might be longer processes or sometimes they might be shorter but knowing these new procedures might gain you some extra credits in your class. Many teachers take these examples as extra credit homework.

Before we can proceed to the special cases, there are some general rules of algebra we will need to review. These rules have specific names, given to them in the algebra world. Yes, their names will be shown in this book, but more importantly is to remember what they are and when can they be applied.

Binomial squared

Every time we have to elements adding each other inside of a parenthesis and the parenthesis is being raised to the power of two, we must use this pattern. Do NOT just distribute the exponent:

$$(A + B)^2 = A^2 + 2 \cdot A \cdot B + B^2$$
$$\cancel{(A + B)^2 \neq A^2 + B^2}$$

Example:
$$(x + 3)^2 = x^2 + 2 \cdot x \cdot 3 + 3^2 = x^2 + 6x + 9$$

Same rule applies when there is a subtraction. The only thing that changes is the first sign which instead of being positive is now negative:

$$(A - B)^2 = A^2 - 2 \cdot A \cdot B + B^2$$
$$\cancel{(A - B)^2 \neq A^2 - B^2}$$

Example:
$$(x - 4)^2 = x^2 - 2 \cdot x \cdot 4 + 2^2 = x^2 - 8x + 4$$

Conjugate multiplication

This rule is easier if we just look at the example right away

$$(A - B)(A + B) = A^2 - B^2$$

Example 1:
$$(x - 5)(x + 5) = x^2 - 5^2 = x^2 - 25$$

Example 2:
$$(x + \sqrt{2})(x - \sqrt{2}) = x^2 - (\sqrt{2})^2 = x^2 - 2$$

Number in red is the operation performed in the current step
Number in blue is the operation performed in the previous step

Problem 131) $\frac{x^2-4}{x+2} = 6 - 3x$

$$\frac{x^2 - 4}{x + 2} = 6 - 3x$$

$$\frac{x^2 - 4}{x + 2} = 6 - 3x$$

$$\boxed{re-write: 4 = 2^2}$$

> There are way too many x's in this problem. Let's see if we can reduce that amount

$$\frac{x^2 - 2^2}{x + 2} = 6 - 3x$$

$$\frac{x^2 - 2^2}{x + 2} = 6 - 3x$$

$$\boxed{Apply\ (A - B)(A + B) = A^2 - B^2}$$

$$\frac{(x - 2)(x + 2)}{x + 2} = 6 - 3x$$

$$\frac{(x - 2)(x + 2)}{x + 2} = 6 - 3x$$

> Well, now we have even more x's but please bear with us one more step

$$\boxed{Cancel\ equal\ terms}$$

$$x - 2 = 6 - 3x$$

$$x - 2 = 6 - 3x$$

$$\boxed{Moves\ as\ positive}$$

> There it is! We eliminated an x and we also got rid of the fraction

$$x + 3x - 2 = 6$$

$$\boxed{Moves\ as\ positive}$$

$$x + 3x = 6 + 2$$

$$4x = 8$$

$$\boxed{Moves\ as\ division}$$

$$x = \frac{8}{4}$$

$$x = 2 \text{ »» } \textbf{Answer}$$

Number in red is the operation performed in the current step
Number in blue is the operation performed in the previous step

Problem 132) $\frac{x^2-7}{x+\sqrt{7}} = 10 - 2x$

$$\frac{x^2-7}{x+\sqrt{7}} = 10 - 2x$$

$$\frac{x^2-7}{x+\sqrt{7}} = 10 - 2x$$

$$re-write: 7 = \left(\sqrt{7}\right)^2$$

Again, too many x's and too many fractions. Let's apply a similar strategy

$$\frac{x^2 - \left(\sqrt{7}\right)^2}{x+\sqrt{7}} = 10 - 2x$$

$$\frac{x^2 - \left(\sqrt{7}\right)^2}{x+\sqrt{7}} = 10 - 2x$$

$$Apply\ (A-B)(A+B) = A^2 - B^2$$

Turning 7 into the square root of 7 squared does not seem like something we may think as the first choice but after enough practice, this idea becomes a very useful resource

$$\frac{\left(x-\sqrt{7}\right)\left(x+\sqrt{7}\right)}{x+\sqrt{7}} = 10 - 2x$$

$$\frac{\left(x-\sqrt{7}\right)\left(x+\sqrt{7}\right)}{x+\sqrt{7}} = 10 - 2x$$

$$x - \sqrt{7} = 10 - 2x$$

$$x - \sqrt{7} = 10 - 2x$$

Moves as positive

$$x + 2x - \sqrt{7} = 10$$

One more time, we have reduced the number of x's and we also eliminated the fraction

Moves as positive

$$x + 2x = 10 + \sqrt{7}$$

$$3x = 10 + \sqrt{7}$$

Moves as division

$$x = \frac{10+\sqrt{7}}{3} \ \text{»»} \ \boldsymbol{x = \frac{10+\sqrt{7}}{3}} \ \text{»» } \boldsymbol{Answer}$$

Number in red is the operation performed in the current step
Number in blue is the operation performed in the previous step

A game of square roots

The next four problems involve square roots in awkward situations. In these cases, we will try to move the roots around, eliminating them by turning them to exponents one by one until we have put an end to all of them. Be aware, we might have to use the rules of algebra we learned at the beginning of this chapter.

Problem 133) $\sqrt{x + 3} = \sqrt{11 - x}$

$$\sqrt{x + 3} = \sqrt{11 - x}$$

$$\sqrt{x + 3} = \sqrt{11 - x}$$

$$\boxed{Apply \ \sqrt{A} = \sqrt[2]{A}}$$

$$\sqrt[2]{x + 3} = \sqrt[2]{11 - x}$$

$$x + 3 = 11 - x$$

$$\boxed{\text{Moves as positive}}$$

$$x + x + 3 = 11$$

$$\boxed{\text{Moves as negative}}$$

$$x + x = 11 - 3$$

$$2x = 8$$

$$\boxed{\text{Moves as division}}$$

$$x = \frac{8}{2}$$

$$x = 4 \ \text{»» } \boldsymbol{Answer}$$

There is a square root on both sides of the equation. We can eliminate them right away. Remember we can do almost whatever we want in an equation as long as we do the same on both sides

Number in red is the operation performed in the current step
Number in blue is the operation performed in the previous step

177

Problem 134) $\sqrt{x^2 + 1} = x - 4$

$$\sqrt{x^2 + 1} = x - 4$$

$$\sqrt[2]{x^2 + 1} = x - 4$$

$$\sqrt[2]{x^2 + 1} = x - 4$$

Moves as an exponent

Every single time a root does not have any number, we can safely assume it is a square root:

$$\sqrt{A} = \sqrt[2]{A}$$

$$x^2 + 1 = (x - 4)^2$$

$$x^2 + 1 = (x - 4)^2$$

Apply $(A - B)^2 = A^2 - 2 \cdot A \cdot B + B^2$

$$x^2 + 1 = x^2 - 2 \cdot x \cdot 4 + 4^2$$

$$x^2 + 1 = x^2 - 8x + 16$$

$$1 = -8x + 16$$

Moves as negative

Negative on the numerator and negative on the denominator of the fraction, this means positive answer

$$1 - 16 = -8x$$

Moves as division

As long as we do exactly the same thing on both sides of the equation, it is fine. So we are allowed to eliminate the x^2 on both sides to make the equation easier

$$\frac{-15}{-8} = x$$

$$x = \frac{15}{8} \quad \text{»» } Answer$$

!

Take a look at the note at the end of problem 136

Problem 135) $\sqrt{x+4} = \sqrt{x+9} - 5$

$$\sqrt{x+4} = \sqrt{x+9} - 5$$

$$\sqrt[2]{x+4} = \sqrt[2]{x+9} - 5$$

| Moves as an exponent |

One more reminder about:

$$\sqrt{A} = \sqrt[2]{A}$$

Even though there is one square root on each side of the equation, that extra -5 won't let us cancel them as we did in problem 133. In these cases, we will move the roots around trying to eliminate them one by one

$$x+4 = \left(\sqrt[2]{x+9} - 5\right)^2$$

$$x+4 = \left(\sqrt[2]{x+9} - 5\right)^2$$

| Apply $(A-B)^2 = A^2 - 2 \cdot A \cdot B + B^2$ |

$$x+4 = \left(\sqrt[2]{x+9}\right)^2 - 2 \cdot \sqrt[2]{x+9} \cdot 5 + 5^2$$

$$x+4 = \left(\sqrt[2]{x+9}\right)^2 - 2 \cdot \sqrt[2]{x+9} \cdot 5 + 5^2$$

The square root cancels with the squared exponent

$$x+4 = x+9 - 10\sqrt[2]{x+9} + 25$$

$$4 = 9 - 10\sqrt[2]{x+9} + 25$$

| Move as negative |

And one more time, we can cancel anything that is repeated on both sides of the equation. Like the x's in this case

$$4 - 9 - 25 = -10\sqrt[2]{x+9}$$

| Moves as division |

$$\frac{-30}{-10} = \sqrt[2]{x+9}$$

| Moves as an exponent |

$$3^2 = x + 9$$

| Move as negative |

$$9 - 9 = x$$

$$x = 0 \text{ »» } \textbf{Answer}$$

!

Take a look at the note at the end of problem 136

Number in red is the operation performed in the current step
Number in blue is the operation performed in the previous step

Problem 136) $\sqrt{x-6} = \sqrt{x-1} + 5$

$$\sqrt{x-6} = \sqrt{x-1} + 5$$

$$\sqrt[2]{x-6} = \sqrt[2]{x-1} + 5$$

Moves as an exponent

Very similar situation as the previous problem, that extra positive 5 won't let us cancel both square roots

$$x - 6 = \left(\sqrt[2]{x-1} + 5\right)^2$$

$$x - 6 = \left(\sqrt[2]{x-1} + 5\right)^2$$

Apply $(A+B)^2 = A^2 + 2 \cdot A \cdot B + B^2$

$$x - 6 = \left(\sqrt[2]{x-1}\right)^2 + 2 \cdot \sqrt[2]{x-1} \cdot 5 + 5^2$$

The square root cancels with the squared exponent

$$x - 6 = \left(\sqrt[2]{x-1}\right)^2 + 2 \cdot \sqrt[2]{x-1} \cdot 5 + 5^2$$

$$x - 6 = x - 1 + 10\sqrt[2]{x-1} + 25$$

Move as positive

Yes, we can still cancel those two x's because they are repeated on both sides of the equation

$$-6 = -1 + 10\sqrt[2]{x-1} + 25$$

Move as negative

$$-6 + 1 - 25 = 10\sqrt[2]{x-1}$$

Moves as division

$$\frac{-30}{10} = \sqrt[2]{x-1}$$

Moves as an exponent

$$(-3)^2 = x - 1$$

Move as positive

$$9 + 1 = x$$

$$x = 10 \;\;»»\; \textbf{\textit{Answer}}$$

Note: Problems 134, 135 and 136 makes us believe they have been solved but if we plug the answers back in the original equation it will not work. There is no mistake, they just DO NOT have real solutions. This is a topic in more advanced math classes but we included them to highlight the importance of plugging the answer back in the original equation to check its validity

Number in red is the operation performed in the current step
Number in blue is the operation performed in the previous step

Disguised Quadratics

Earlier in this book we discussed about quadratic equations. We talked about them being a different type of equations that can't be solved using the "opposite rule" to move things around the equation to isolate the x, instead we were supposed to use either the quadratic formula or factorization to solve. Sometimes we will face equations that may not appear quadratic at first sight but if we pay attention to the details, we will see they are actually a quadratic equation in disguise. First let's recall the generic quadratic equation:

$$Ax^2 + Bx + C = 0$$

There is a term with an x squared, a term with an x not squared and a term without an x. All that equal to zero. Any equation following this pattern is considered a quadratic equation. The following are examples of quadratic equations:

$$x^2 + 4x - 10 = 0$$

$$5x^2 - 3x - 8 = 0$$

Watch carefully the next equation:

$$x^6 + 2x^3 - 1 = 0$$

Our first thought may not consider this a quadratic equation, the fact of having an x to the power of 6 is the first flag. But take a look to the exponents of the x:

$$x^6 + 2x^3 - 1 = 0$$

The number of the highest exponent (the six in purple or dark) is twice as big as the exponent of the middle element (the 3 in green or light). If we apply the rule of $(A^M)^N = A^{M \cdot N}$ we could re-write this equation as:

$$(x^2)^3 + 2(x)^3 - 1 = 0$$

Looking at the numbers in the parenthesis, it is now a quadratic equation but that power of 3 (in blue) is not letting us finish the problem, but we could do a little algebra trick. Recall the original problem:

$$x^6 + 2x^3 - 1 = 0$$

Take the part in red of the middle term and set it equal to a different variable, any variable:
$$x^3 = y$$
Therefore, based on exponent rules:
$$x^6 = (x^3)^2 = y^2$$
Now the equation can be re-written as:
$$y^2 + 2y + 1 = 0$$

Now we have a quadratic equation. The next four problems will show all the steps about how to set up disguised quadratics and how to solve them.

Number in red is the operation performed in the current step
Number in blue is the operation performed in the previous step

181

Problem 137) $x^4 + 3x^2 - 10 = 0$

$$x^4 + 3x^2 - 10 = 0$$

$$x^4 + 3x^2 - 10 = 0$$

Check if the highest exponent is twice the middle term

It is not a quadratic equation since the highest exponent of x is 4, but let's check if it is a disguised quadratic

Dividing the highest exponent by the exponent of the middle term is a good check. If the answer is equal to two, then it is twice as big

$$\frac{4}{2} = 2$$

$$x^4 + 3x^2 - 10 = 0$$

Set the middle term equal to a different variable

$$x^2 = y$$

By definition the highest term is now:

$$x^4 = y^2$$

The division is equal to two, it is twice as big therefore, this is a disguised quadratic

If the earlier division of the exponents equals 2, then this situation will always happen:
$middle\ term = y$
$highest\ x = y^2$

$$x^4 + 3x^2 - 10 = 0$$

Replace the x's for the new variables

$$y^2 + 3y - 10 = 0$$

$$1y^2 + 3y - 10 = 0$$

$$A = 1 \; ; B = 3 \; ; C = -10$$

Identify the values for A, B and C

$$\frac{-B \pm \sqrt{B^2 - 4 \cdot A \cdot C}}{2 \cdot A}$$

$$\frac{-(3) \pm \sqrt{(3)^2 - 4 \cdot (1) \cdot (-10)}}{2 \cdot 1}$$

Plug the values into the formula

$$= \frac{-3 \pm \sqrt{9 + 40}}{2}$$

$$= \frac{-3 \pm \sqrt{49}}{2}$$

Number in red is the operation performed in the current step
Number in blue is the operation performed in the previous step

$$= \frac{-3 \pm 7}{2}$$

Split the plus or minus sign to get the two answers

$$y = \frac{-3 + 7}{2}$$

$$y = \frac{4}{2}$$

$$y = 2$$

$$y = \frac{-3 - 7}{2}$$

$$y = \frac{-10}{2}$$

$$y = -5$$

The problem is not over yet; we were given a problem in terms of x but we have an answer in terms of y. We must bring the answers back to the initial variable

$$y = 2 \ \ and \ \ y = -5$$

Get the answer back into terms of x

$$If \ x^2 = y \ then \ x^2 = 2$$

$$x^2 = 2$$

Moves as root

$$x = \sqrt{2}$$

$$x = \pm\sqrt{2}$$

$$If \ x^2 = y \ then \ x^2 = -5$$

$$x^2 = -5$$

Moves as root

$$x = \sqrt{-5}$$

Negative numbers do not have square root. No answer on this side

$$x = \sqrt{2} \ \ and \ \ x = -\sqrt{2} \ \text{»» } \textbf{\textit{Answer}}$$

Number in red is the operation performed in the current step
Number in blue is the operation performed in the previous step

Problem 138) $x - 5\sqrt{x} + 6 = 0$

$$x - 5\sqrt{x} + 6 = 0$$

$$x - 5\sqrt{x} + 6 = 0$$

$$\boxed{Apply \ \sqrt{A} = A^{\frac{1}{2}}}$$

$$x - 5x^{\frac{1}{2}} + 6 = 0$$

At first sight is not a quadratic. Let's check for disguised quadratics

$$x^1 - 5x^{\frac{1}{2}} + 6 = 0$$

$$\boxed{\text{Check if the highest exponent is twice the middle term}}$$

When a number or variable does not have a written exponent we can assume its exponent is equal to 1

$$\frac{1}{1/2} = \frac{1 \cdot 2}{1} = 2$$

$$x - 5x^{\frac{1}{2}} + 6 = 0$$

$$\boxed{\text{Set the middle term equal to a different variable}}$$

$$x^{\frac{1}{2}} = y$$

The division is equal to two, it is twice as big therefore, this is a disguised quadratic

By definition the highest term is now:

$$x = y^2$$

If the earlier division of the exponents equals 2, then this situation will always happen:

$$middle \ term = y$$
$$highest \ x = y^2$$

$$x - 5x^{\frac{1}{2}} + 6 = 0$$

$$\boxed{\text{Replace the x's for the new variables}}$$

$$y^2 - 5y + 6 = 0$$

There it is, a quadratic equation!

$$1y^2 - 5y + 6 = 0$$

$$A = 1 \ ; B = -5 \ ; C = 6$$

$$\boxed{\text{Identify the values for A, B and C}}$$

$$\frac{-B \pm \sqrt{B^2 - 4 \cdot A \cdot C}}{2 \cdot A}$$

$$\frac{-(-5) \pm \sqrt{(-5)^2 - 4 \cdot (1) \cdot (6)}}{2 \cdot 1}$$

$$\boxed{\text{Plug the values into the formula}}$$

Number in red is the operation performed in the current step

Number in blue is the operation performed in the previous step

$$= \frac{5 \pm \sqrt{25 - 24}}{2}$$

$$= \frac{5 \pm \sqrt{1}}{2}$$

$$= \frac{5 \pm 1}{2}$$

> Split the plus or minus sign to get the two answers

$$y = \frac{5 + 1}{2} \qquad y = \frac{5 - 1}{2}$$

$$y = \frac{6}{2} \qquad y = \frac{4}{2}$$

$$y = 3 \qquad y = 2$$

$$y = 3 \quad and \quad y = 2$$

The problem is not over yet; we were given a problem in terms of x but we have an answer in terms of y. We must bring the answers back to the initial variable

> Get the answer back into terms of x

Remember $x^{\frac{1}{2}} = \sqrt[2]{x}$

$$If \ \sqrt[2]{x} = y \ then \ \sqrt[2]{x} = 3 \qquad\qquad If \ \sqrt[2]{x} = y \ then \ \sqrt[2]{x} = 2$$

$$\sqrt[2]{x} = 3 \qquad\qquad\qquad \sqrt[2]{x} = 2$$

> Moves as an exponent

> Moves as an exponent

$$x = 3^2 \qquad\qquad\qquad x = 2^2$$

$$x = 9 \qquad\qquad\qquad\qquad x = 4$$

$$x = 9 \quad and \quad x = 4 \ \text{»» } Answer$$

Number in red is the operation performed in the current step
Number in blue is the operation performed in the previous step

185

Problem 139) $2x^{\frac{2}{3}} - 7x^{\frac{1}{3}} + 6 = 0$

$$2x^{\frac{2}{3}} - 7x^{\frac{1}{3}} + 6 = 0$$

$$2x^{\frac{2}{3}} - 7x^{\frac{1}{3}} + 6 = 0$$

Check if the highest exponent is twice the middle term

At first sight is not a quadratic. Let's check for disguised quadratics

$$\frac{^2/_3}{^1/_3} = \frac{2 \cdot 3}{1 \cdot 3} = 2$$

For double fractions apply the rule:

$$\frac{\frac{A}{B}}{\frac{C}{D}} = \frac{A \cdot D}{C \cdot D} \rightarrow \frac{\frac{2}{3}}{\frac{1}{3}} = \frac{2 \cdot 3}{1 \cdot 3} = \frac{6}{3}$$

$$= 2$$

$$x^{\frac{2}{3}} - 7x^{\frac{1}{3}} + 6 = 0$$

Set the middle term equal to a different variable

$$x^{\frac{1}{3}} = y$$

The division is equal to two, it is twice as big therefore, this is a disguised quadratic

By definition the highest term is now:

$$x^{\frac{2}{3}} = y^2$$

$$2x^{\frac{2}{3}} - 7x^{\frac{1}{3}} + 6 = 0$$

Replace the x's for the new variables

There it is, a quadratic equation!

If the earlier division of the exponents equals 2, then this situation will always happen:

$$middle\ term = y$$
$$highest\ x = y^2$$

$$2y^2 - 7y + 6 = 0$$

$$2y^2 - 7y + 6 = 0$$

$$A = 2\ ; B = -7\ ; C = 6$$

Identify the values for A, B and C

$$\frac{-B \pm \sqrt{B^2 - 4 \cdot A \cdot C}}{2 \cdot A}$$

$$\frac{-(-7) \pm \sqrt{(-7)^2 - 4 \cdot (2) \cdot (6)}}{2 \cdot 2}$$

Plug the values into the formula

$$= \frac{7 \pm \sqrt{49 - 48}}{4}$$

Number in red is the operation performed in the current step
Number in blue is the operation performed in the previous step

$$= \frac{7 \pm \sqrt{1}}{4}$$

$$= \frac{7 \pm 1}{4}$$

Split the plus or minus sign to get the two answers

$$y = \frac{7 + 1}{4} \qquad\qquad y = \frac{7 - 1}{4}$$

$$y = \frac{8}{4} \qquad\qquad y = \frac{6}{4}$$

$$y = 2 \qquad\qquad y = \frac{3}{2}$$

The problem is not over yet; we were given a problem in terms of x but we have an answer in term. We must bring the answers back to the initial variable

$$y = 2 \;\; and \;\; y = \frac{3}{2}$$

Get the answer back into terms of x

Remember $x^{\frac{1}{3}} = \sqrt[3]{x}$

$$If \; \sqrt[3]{x} = y \; then \; \sqrt[3]{x} = 2 \qquad\qquad If \; \sqrt[3]{x} = y \; then \; \sqrt[3]{x} = \frac{3}{2}$$

$$\sqrt[3]{x} = 2 \qquad\qquad\qquad \sqrt[3]{x} = \frac{3}{2}$$

Moves as an exponent	Moves as an exponent

$$x = 2^3 \qquad\qquad\qquad x = \left(\frac{3}{2}\right)^3$$

$$x = 8 \qquad\qquad\qquad x = \frac{27}{8}$$

$$x = 8 \; and \; x = \frac{27}{8} \;\; »» \; Answer$$

Number in red is the operation performed in the current step
Number in blue is the operation performed in the previous step

Problem 140) $\frac{\sqrt{12x-5}}{x} = 2$

$$\frac{\sqrt{12x - 5}}{x} = 2$$

Does not look like a quadratic, but let's re-arrange the problem and check

$$\frac{\sqrt{12x - 5}}{x} = 2$$

Moves as multiplication

$$\sqrt{12x - 5} = 2 \cdot x$$

Apply $\sqrt{A} = \sqrt[2]{A}$

$$\sqrt[2]{12x - 5} = 2x$$

$$\sqrt[2]{12x - 5} = 2x$$

Moves as an exponent

$$12x - 5 = (2x)^2$$

$$12x - 5 = (2x)^2$$

Apply: $(A \cdot B)^M = A^M \cdot B^M$

$$12x - 5 = 2^2 \cdot x^2$$

$$12x - 5 = 4x^2$$

Moves as negative

Put everything on the same side of the equation. We strongly recommend to move everything to the side where the positive x^2 is located

$$-5 = 4x^2 - 12x$$

Moves as positive

$$0 = 4x^2 - 12x + 5$$

$$0 = 4x^2 - 12x + 5$$

This is a quadratic equation!

Number in red is the operation performed in the current step

Number in blue is the operation performed in the previous step

$$0 = 4x^2 - 12x + 5$$

$$A = 4 \; ; B = -12 \; ; C = 5$$

Identify the values for A, B and C

$$\frac{-B \pm \sqrt{B^2 - 4 \cdot A \cdot C}}{2 \cdot A}$$

$$= \frac{-(-12) \pm \sqrt{(-12)^2 - 4 \cdot (4) \cdot (5)}}{2 \cdot 4}$$

Plug the values into the formula

$$= \frac{12 \pm \sqrt{144 - 80}}{8}$$

$$= \frac{12 \pm \sqrt{64}}{8}$$

$$= \frac{12 \pm 8}{8}$$

Split the plus or minus sign to get the two answers

$$x = \frac{12 + 8}{8} \qquad\qquad x = \frac{12 - 8}{8}$$

$$x = \frac{20}{8} \qquad\qquad x = \frac{4}{8}$$

$$x = \frac{5}{2} \qquad\qquad x = \frac{1}{2}$$

Always simplify fractions:

$$\frac{20 \div 4}{8 \div 4} = \frac{5}{2}$$

Always simplify fractions:

$$\frac{4 \div 4}{8 \div 4} = \frac{1}{2}$$

$$x = \frac{5}{2} \;\; and \;\; x = \frac{1}{2} \;\; \text{»» Answer}$$

Disguised quadratics: practice problems

1) $x + 7\sqrt{x} - 18 = 0$ 2) $x^8 - 2x^4 - 3 = 0$

3) $x^{\frac{2}{3}} - 4x^{\frac{1}{3}} - 5 = 0$ 4) $e^{2x} + 3e^x + 2 = 0$

Number in red is the operation performed in the current step
Number in blue is the operation performed in the previous step

Completing the square

Before we even start talking about *completing the square*, we are obligated to say; most of the problems that can be solved using this technique can also be solved using *factorization* and *the quadratic formula*. There are two reasons we decided to include *completing the square* in this book: first, it is a technique that will be useful in more advance math like differential calculus, second, there are really good possibilities solving a *completing the square* problem in a homework or an exam will give you extra credit.

Our main goal of this section will be to force our equations to look like the binomial square rule:

$$(A + B)^2 = A^2 + 2 \cdot A \cdot B + B^2$$

Or its subtracting counterpart

$$(A - B)^2 = A^2 - 2 \cdot A \cdot B + B^2$$

So for example look at the following binomial square:

$$(x + 4)^2 = x^2 + 2 \cdot x \cdot 4 + 4^2 = x^2 + 8x + 16$$

Now let's take a look at the following equation

$$x^2 + 8x + 10$$

It looks very similar to the binomial square shown before but we have a +10 instead of a +16

If we decide to add a +6 to the equation

$$x^2 + 8x + 10 \; +6$$

Now we get

$$x^2 + 8x + 16 = (x + 4)^2$$

This is the principle we will follow to solve problems using completing the square. Of course in each example we will show you all the steps that need to be made and rules that apply on each different situation.

Number in red is the operation performed in the current step
Number in blue is the operation performed in the previous step

Problem 141) $x^2 + 6x + 12 = 19$

$$x^2 + 6x + 12 = 19$$

$$x^2 + 6x + 12 = 19$$

Separate the two elements with x

This step is not strictly necessary. It is just a reminder we are trying to resemble the binomial square rule

$$x^2 + 6x$$

$$x^2 + 6x + B^2 = A^2 + 2 \cdot A \cdot B + B^2$$

$$x^2 + 6x + B^2$$

Divide the number multiplying the x^1 by 2

The only thing we are missing to complete our binomial is the "B" part. Dividing by 2 the number with the x (the one in red) will always be the key to find it. ALWAYS!

$$\frac{6}{2} = 3$$

$$B = 3 \rightarrow therefore \rightarrow B^2 = 3^2 = 9$$

$$B^2 = 9$$

Add the new found number to BOTH sides of the ORIGINAL equation

Always, squaring the answer of this division will give us the missing element to have our binomial square

$$x^2 + 6x + 9 + 12 = 19 + 9$$

$$x^2 + 6x + 9 + 12 = 28$$

Identify the elements of the binomial square

Remember we can do almost whatever we want to an equation as long as we do it to both sides of the equal sign

$$A^2 = x^2 \; ; 2 \cdot A \cdot B = 2 \cdot x \cdot 3 = 6x \; ; B^2 = 9$$

$$A^2 + 2AB + B^2 = (A + B)^2$$

$$x^2 + 6x + 9 = (x + 3)^2$$

Replace terms on the problem

The number we get from the division by 2 a few steps ago will be our B value in most cases

$$(x + 3)^2 + 12 = 28$$

Moves as negative

$$(x + 3)^2 = 28 - 12$$

Moves as root

Number in red is the operation performed in the current step
Number in blue is the operation performed in the previous step

$$x + 3 = \sqrt[2]{16}$$

Account for both positive and
negative answers of the root

$x + 3 = 4$	$x + 3 = -4$
Moves as negative	Moves as negative
$x = 4 - 3$	$x = -4 - 3$
$x = 1$	$x = -7$

$$x = 1 \;\; and \;\; x = -7 \;\text{»» } \boldsymbol{Answers}$$

Number in red is the operation performed in the current step
Number in blue is the operation performed in the previous step

Problem 142) $x^2 - 10x + 7 = 63$

$$x^2 - 10x + 7 = 63$$

$$x^2 - 10x + 7 = 63$$

Separate the two elements with x

$$x^2 - 10x$$

$$x^2 - 10x + B^2 = A^2 + 2 \cdot A \cdot B + B^2$$

$$x^2 - 10x + B^2$$

Divide the number multiplying the x^1 by 2

As problem 141, we are resembling the binomial square, but this time is a subtraction

Yes, repeat all the process again. In this step we are supposed to take the number multiplying the x and divide it by two. It is the key!

$$\frac{10}{2} = 5$$

$$B = 5 \rightarrow therefore \rightarrow B^2 = 5^2 = 25$$

$$B^2 = 25$$

Add the new found number to BOTH sides of the ORIGINAL equation

And now, we have to square the answer of the division to find our missing term to complete the binomial

$$x^2 - 10x + 25 + 7 = 63 + 25$$

$$x^2 - 10x + 25 + 7 = 88$$

Identify the elements of the binomial square

Remember we can do almost whatever we want to an equation as long as we do it to both sides of the equal sign

$$A^2 = x^2 \; ; 2 \cdot A \cdot B = 2 \cdot x \cdot 5 = 10x \; ; B^2 = 25$$

$$A^2 - 2AB + B^2 = (A - B)^2$$

$$x^2 - 10x + 25 = (x - 5)^2$$

Replace terms on the problem

The number we get from the division by 2 a few steps ago will be our B value

$$(x - 5)^2 + 7 = 88$$

Moves as negative

$$(x - 5)^2 = 88 - 7$$

Moves as root

Number in red is the operation performed in the current step
Number in blue is the operation performed in the previous step

$$x - 5 = \sqrt[2]{81}$$

Account for both positive and
negative answers of the root

$$x - 5 = 9 \qquad\qquad x - 5 = -9$$

Moves as positive | Moves as positive

$$x = 9 + 5 \qquad\qquad x = -9 + 5$$

$$x = 14 \qquad\qquad x = -4$$

$$x = 14 \ and \ x = -4 \ \text{»» } Answers$$

Problem 143) $4x^2 - 4x + 5 = 8$

$$4x^2 - 4x + 5 = 8$$

$$4x^2 - 4x + 5 = 8$$

This time, the x² is not alone. Few things change when this situation occurs

Separate the two elements with x

$$4x^2 - 4x$$

This is still our main objective but now we also have to take a closer look to the A²

$$4x^2 - 4x + B^2 = A^2 - 2 \cdot A \cdot B + B^2$$

Take the elements in red

$$4x^2 = A^2$$

$$2x = A$$

Playing with algebra here:
if $4x^2 = A^2$
Take the square root on both sides
Then $\sqrt[2]{4x^2} = \sqrt[2]{A^2}$
$\sqrt{4} \cdot \sqrt{x^2} = \sqrt{A^2}$
$2x = A$

$$4x^2 - 4x + B^2 = A^2 - 2 \cdot A \cdot B + B^2$$

Take the middle term

$$-4x = -2 \cdot A \cdot B$$

We need to solve for B here

$$-4x = -2 \cdot A \cdot B$$

Moves as division

$$\frac{-4x}{-2} = A \cdot B$$

Moves as division

$$\frac{2x}{A} = B$$

Remember

A=2x

$$\frac{2x}{2x} = B$$

$$1 = B$$

Now that we've finally found B² we can proceed with completing the square

$$Then\ B^2 = 1^2 = 1$$

Add the new found number to BOTH sides of the ORIGINAL equation

$$4x^2 - 4x + 1 + 5 = 8 + 1$$

Number in red is the operation performed in the current step
Number in blue is the operation performed in the previous step

$$4x^2 - 4x + 1 + 5 = 9$$

Identify the elements of the binomial square

$$A^2 = 4x^2 \;;\; 2 \cdot A \cdot B = 2 \cdot (2x) \cdot (1) = 4x \;;\; B^2 = 1$$

$$A^2 - 2AB + B^2 = (A - B)^2$$

$$4x^2 - 4x + 1 = (2x - 1)^2$$

Replace terms on the problem

$$(2x - 1)^2 + 5 = 9$$

Moves as negative

$$(2x - 1)^2 = 9 - 5$$

Moves as root

$$2x - 1 = \sqrt[2]{4}$$

Account for both positive and negative answers of the root

$2x - 1 = 2$	$2x - 1 = -2$
Moves as positive	Moves as positive
$2x = 2 + 1$	$2x = -2 + 1$
Moves as division	Moves as division
$x = \dfrac{3}{2}$	$x = \dfrac{-1}{2}$
$x = \dfrac{3}{2}$	$x = -\dfrac{1}{2}$

$$x = \frac{3}{2} \;\; and \;\; x = -\frac{1}{2} \;\; \text{»» } \textbf{\textit{Answers}}$$

Number in red is the operation performed in the current step
Number in blue is the operation performed in the previous step

Problem 144) $x^2 + 9x + 1 = 4$

$$x^2 + 9x + 1 = 4$$

$$x^2 + 9x + 1 = 4$$

Separate the two elements with x

This is our main objective, don't forget about it!

$$x^2 + 9x$$

$$x^2 + 9x + B^2 = A^2 + 2 \cdot A \cdot B + B^2$$

$$x^2 + 9x + B^2$$

Divide the number multiplying the x^1 by 2

Whatever number is multiplying the x without exponent gets divided by two. That is the game

$$\frac{9}{2} = 4.5$$

$$B = \frac{9}{2} \rightarrow therefore \rightarrow B^2 = \frac{9^2}{2^2} = \frac{81}{4}$$

Most Algebra courses and teachers will want you to keep your answer as a fraction, but decimal is also fine

$$B^2 = \frac{81}{4}$$

Add the new found number to BOTH sides of the ORIGINAL equation

Remember we can do almost whatever we want to an equation as long as we do it to both sides of the equal sign

$$x^2 + 9x + \frac{81}{4} + 1 = 4 + \frac{81}{4}$$

$$x^2 + 9x + \frac{81}{4} + 1 = \frac{97}{4}$$

Identify the elements of the binomial square

Apply:
$$\frac{A}{B} + \frac{C}{D} = \frac{A \cdot D + B \cdot C}{B \cdot D}$$
$$\frac{4}{1} + \frac{81}{4} = \frac{4 \cdot 4 + 81 \cdot 1}{1 \cdot 4}$$
$$= \frac{16 + 81}{4} = \frac{97}{4}$$

$$A^2 = x^2 \; ; 2 \cdot A \cdot B = 2 \cdot x \cdot \frac{9}{2} = 9x \; ; B^2 = \frac{81}{4}$$

$$A^2 + 2AB + B^2 = (A + B)^2$$

$$x^2 + 9x + \frac{81}{4} = \left(x + \frac{9}{2} \right)^2$$

Replace terms on the problem

Number in red is the operation performed in the current step

Number in blue is the operation performed in the previous step

$$\left(x + \frac{9}{2}\right)^2 + 1 = \frac{97}{4}$$

| Moves as negative |

$$\left(x + \frac{9}{2}\right)^2 = \frac{97}{4} - 1$$

| Moves as root |

Apply:

$$\frac{A}{B} - \frac{C}{D} = \frac{A \cdot D - B \cdot C}{B \cdot D}$$

$$\frac{97}{4} - \frac{1}{1} = \frac{97 \cdot 1 - 4 \cdot 1}{4 \cdot 1}$$

$$= \frac{97 - 4}{4} = \frac{93}{4}$$

$$x + \frac{9}{2} = \sqrt[2]{\frac{93}{4}}$$

Roots and exponents distribute through fractions:

$$\sqrt{\frac{A}{B}} = \frac{\sqrt{A}}{\sqrt{B}}$$

Don't forget square roots have both positive and negative answers

$$x + \frac{9}{2} = \pm\frac{\sqrt{93}}{\sqrt{4}}$$

$$x + \frac{9}{2} = \pm\frac{\sqrt{93}}{2}$$

| Account for both positive and negative answers of the root |

When both fractions have the same denominator, apply this:

$$\frac{A}{C} + \frac{B}{C} = \frac{A + B}{C}$$

$$x + \frac{9}{2} = \frac{\sqrt{93}}{2} \qquad\qquad x + \frac{9}{2} = -\frac{\sqrt{93}}{2}$$

| Moves as negative | | Moves as negative |

$$x = \frac{\sqrt{93}}{2} - \frac{9}{2} \qquad\qquad x = -\frac{\sqrt{93}}{2} - \frac{9}{2}$$

$$x = \frac{\sqrt{93} - 9}{2} \qquad\qquad x = \frac{-\sqrt{93} - 9}{2}$$

$$x = \frac{\sqrt{93} - 9}{2} \quad and \quad x = \frac{-\sqrt{93} - 9}{2} \quad \text{»» } \textbf{\textit{Answers}}$$

Completing the square: practice problems

1) $x^2 + 6x + 1 = 3$ 2) $x^2 + 7x - 5 = 2$
3) $x^2 + 1 = 8x - 5$ 4) $4x^2 - 1 = 5 - 2x$

Number in red is the operation performed in the current step
Number in blue is the operation performed in the previous step

Systems of two or more variables

We've reached the final section of this book. If we are still considering math as a language and we think of Algebra as a dialect, then this is the section in which we learn how to talk about politics and history with native speakers. So far we have learned a large amount of techniques to solve for many different types of equations and situations we may encounter, but what happens when our problem has more than just one variable? We are not talking about the same variable being repeated many times in the problem, we did that already! We are talking about problems involving different variables, like having x and y in the same problem!

Actually there is nothing to worry about, if you feel comfortable with all the techniques learned before, this is just an extra step.

We need to highlight not all *systems of two or more variables* can be solved, which is why we first need to check the following rule before we attempt any solution:

<u>**A system of equations can be solved using algebra only if the number of variables is equal or less than the number of equation**</u>

In other words, before we start solving a problem we will count the number of variables involved in the problem, if that number is larger than the number of equations within the same problem, then it CAN NOT be solved. If the number of variables is lesser or equal than the number of equations, then we can proceed.

Once we have checked if the problem meets the criteria our main objective will be to have one final equation involving only one variable. Afterwards, we will solve for that variable and use it to find the others. We will start by choosing any of the equations and solve for one of the variables, anyone. The solution will give us an equation in term of the remaining variables. We'll take that solution and plug it into the next equation and repeat the process until we have reached the final equation with only one variable...

...That sounds so confusing! But it will make a lot more sense with the examples. Trust us!

Number in red is the operation performed in the current step
Number in blue is the operation performed in the previous step

Problem 145) $Eq.1 \quad x + y = 8$
$Eq.2 \quad x - y = 2$

$Eq.1 \quad x + y = 8$
$Eq.2 \quad x - y = 2$

Count the number of variables

Don't count how many variables there are, just how many different ones. Only x and y here

$Number\ of\ variables = 2$
$Number\ of\ equations = 2$

$Number\ of\ equations \geq Number\ of\ variables$

The problem can be solved

Checking if the problem can be solved should always be your first step

$Eq.1 \quad x + y = 8$

$Eq.2 \quad x - y = 2$

Pick any of the two equations

It doesn't matter which one you pick. Try to pick the one that looks easier for you

$x - y = 2$

$x - y = 2$

Solve for any of the two variables

Again, it doesn't matter which one you pick. X was just our initial preference

$x - y = 2$

Moves as positive

We have an equation of x in terms of y. This is our first milestone

$x = 2 + y$

$Solution\ 1 \rightarrow x = 2 + y$

Pick the remaining equation

$Eq.1 \quad x + y = 8$

$x + y = 8$

We did it, we have one equation with only one variable. We know how to solve this!

Here is where we replace x for an equation in terms of y so now we only have one variable to deal with

Replace the variable in red for the equation in Solution 1

$2 + y + y = 8$

$2 + y + y = 8$

Number in red is the operation performed in the current step
Number in blue is the operation performed in the previous step

$$2 + y + y = 8$$

Combine elements with y

$$2 + 2y = 8$$

Moves as negative

$$2y = 8 - 2$$

Moves as division

$$y = \frac{6}{2}$$

$$Solution\ 2 \rightarrow y = 3$$

Recall Solution 1

$$x = 2 + y$$

Replace the variable in red for the value in Solution 2

$$x = 2 + 3$$

$$Solution\ 3 \rightarrow x = 5$$

$$\textbf{\textit{x = 5 and y = 3 »» Answers}}$$

There we go, we have one of the variables. Now let's use it to solve for the other one

Number in red is the operation performed in the current step
Number in blue is the operation performed in the previous step

201

Problem 146) $Eq.\,1 \quad 3x + 2y = 1$
$Eq.\,2 \quad 2x + y = 4$

$Eq.\,1 \quad 3x + 2y = 1$

$Eq.\,2 \quad 2x + y = 4$

Count the number of variables

There are only two different variables, x and y

$Number\ of\ variables = 2$

$Number\ of\ equations = 2$

$Number\ of\ equations \geq Number\ of\ variables$

The problem can be solved

Check if the problem can be solved at first, otherwise you might waste valuable time

$Eq.\,1 \quad 3x + 2y = 10$

$Eq.\,2 \quad 2x + y = 4$

Pick any of the two equations

It really doesn't matter which equation you choose to solve first

$2x + y = 4$

$2x + y = 4$

Solve for any of the two variables

Pick whichever variable looks easier for you. It doesn't matter which one you pick first

$2x + y = 4$

Moves as negative

$y = 4 - 2x$

$Solution\ 1 \rightarrow y = 4 - 2x$

First milestone reached. An equation of y in terms of x

Pick the remaining equation

$Eq.\,1 \quad 3x + 2y = 10$

We did it, we have one equation with only one variable. We know how to solve this!

$3x + 2y = 8$

Replace the variable in red for the equation in Solution 1

$3x + 2(4 - 2x) = 10$

$3x + 2(4 - 2x) = 10$

Number in red is the operation performed in the current step
Number in blue is the operation performed in the previous step

$$3x + 2(4 - 2x) = 10$$

Distribute the 2 inside of the parenthesis

$$3x + 2 \cdot 4 - 2 \cdot 2x = 10$$

$$3x + 8 - 4x = 10$$

Combine elements with x

$$8 - x = 10$$

Moves as negative

$$-x = 10 - 8$$

$$-x = 2$$

Move the negative to the other side

As we have done in previous problems, negatives can be moved from one side to the other in order to solve for x

$$y = -2$$

$$Solution\ 2 \rightarrow y = -2$$

Recall Solution 1

We have a solution for one of the variables, we are half way done

$$y = 4 - 2x$$

Replace the variable in red for the value in Solution 2

$$y = 4 - 2(-2)$$

Careful with the double negative!

$$y = 4 + 4$$

$$Solution\ 3 \rightarrow y = 8$$

$$x = -2\ and\ y = 8\ »»\ Answers$$

Number in red is the operation performed in the current step
Number in blue is the operation performed in the previous step

Problem 147) $Eq.1 \quad 2x - 5y = 4$

$\qquad Eq.2 \quad 4x - 15y = 1$

$Eq.1 \quad 2x - 5y = 4$

$Eq.2 \quad 4x - 15y = 1$

Count the number of variables

There are only two different variables, x and y

$Number\ of\ variables = 2$

$Number\ of\ equations = 2$

$Number\ of\ equations \geq Number\ of\ variables$

The problem can be solved

Start by doing this check!

$Eq.1 \quad 2x - 5y = 4$

$Eq.2 \quad 4x - 15y = 1$

Pick any of the two equations

Just a reminder, pick any of the two equations first, it will give you the same final answer regardless of your choice

$2x - 5y = 4$

$2x - 5y = 4$

Same as the equations, you can pick any of the variables to start with

Solve for any of the two variables

$2x - 5y = 4$

Moves as positive

$2x = 4 + 5y$

Moves as division

$x = \dfrac{4 + 5y}{2}$

Apply:

$$\frac{A + B}{C} = \frac{A}{C} + \frac{B}{C}$$

$$\frac{4 + 5x}{2} = \frac{4}{2} + \frac{5x}{2}$$

$$= 2 + \frac{5x}{2}$$

$x = 2 + \dfrac{5y}{2}$

First milestone reached. An equation of x in terms of y

$Solution\ 1 \rightarrow x = 2 + \dfrac{5y}{2}$

Pick the remaining equation

$Eq.2 \quad 4x - 15y = 1$

Number in red is the operation performed in the current step

Number in blue is the operation performed in the previous step

$$4x - 15y = 1$$

Replace the variable in red for the equation in Solution 1

Now we are replacing x for an equation with just y

$$4\left(2 + \frac{5y}{2}\right) - 15y = 1$$

$$4\left(2 + {}^{5y}/_2\right) - 15y = 1$$

Distribute the 4 inside of the parenthesis

$$4 \cdot 2 + 4 \cdot {}^{5y}/_2 - 15y = 1$$

$$8 + 10y - 15y = 10$$

Combine elements with y

Little help with the fraction multiplication:

$$4 \cdot \frac{5y}{2} = \frac{4}{1} \cdot \frac{5y}{2} = \frac{4 \cdot 5y}{1 \cdot 2}$$
$$= \frac{20y}{2} = 10y$$

$$8 - 5y = 1$$

Moves as negative

$$-5y = 1 - 8$$

Moves as division

Solution 1 is actually:
$$x = 2 + \frac{5y}{2}$$
But a fraction multiplying a variable can be written by separate:
$$x = 2 + \frac{5}{2} \cdot y$$

$$y = {}^{-7}/_{-5}$$

$$Solution\ 2 \rightarrow y = {}^7/_5$$

Recall Solution 1

$$x = 2 + \frac{5}{2} \cdot y$$

Replace the variable in red for the value in Solution 2

Nobody likes fractions but it is a solution for one of the variables so we are good!

$$x = 2 + \frac{5}{2} \cdot \frac{7}{5}$$

$$x = 2 + {}^7/_2$$

$$Solution\ 3 \rightarrow x = {}^{11}/_2$$

$$\frac{2}{1} + \frac{7}{2} = \frac{2 \cdot 2 + 1 \cdot 7}{1 \cdot 2}$$
$$= \frac{4 + 7}{2} = \frac{11}{2}$$

$$x = \frac{11}{2}\ \ and\ \ y = \frac{7}{5}\ \text{»» } Answers$$

Number in red is the operation performed in the current step
Number in blue is the operation performed in the previous step

Problem 148)

$Eq.1 \quad x + 2y - z = 10$

$Eq.2 \quad 5x + z = -5$

$Eq.1 \quad x + 2y - z = 10$

$Eq.2 \quad 5x + z = -5$

Count the number of variables

There are three different variables: x, y and z

$Number\ of\ variables = 3$

$Number\ of\ equations = 2$

$Number\ of\ equations \leq Number\ of\ variables$

The variables outnumber the equations. We have three different and only two equations. This system of equations is breaking the main rule of systems. It can NOT be solved using algebra.

No Answer

Problem 149) $Eq.1 \quad 3x - 2y = 6$

$\qquad Eq.2 \quad 9x - 6y = -5$

$Eq.1 \quad 3x - 2y = 6$

$Eq.2 \quad 9x - 6y = -5$

Actually, this is a tricky question. Even though it meets the criteria for the relationship between the amount of variables and equations, this problem can't be solved.

If you try solving this problem, you will reach a point in which you can't do anything else and still get no answer. The reason this problem can't be solve is because the two equations represent two parallel lines. The solutions for systems of equations represent intersection points and parallel lines do not intersect each other. One of the methods to figure out if the lines are parallel is graphing them. For the other one, take a look at the coefficients of the variables:

$Eq.1 \quad 3x - 2y = 6$

$Eq.2 \quad 9x - 6y = -5$

$Coefficients\ of\ x \rightarrow 3\ and\ 9$

$Divide\ them\ each\ other \rightarrow \dfrac{9}{3} = 3$

$Coefficients\ of\ y \rightarrow -2\ and -6$

$Divide\ them\ each\ other \rightarrow \dfrac{-6}{-2} = 3$

Every time the division of the coefficients are equal to each other, including the same sign, the lines are parallel so the problem has no solution.

No Answer

Number in red is the operation performed in the current step
Number in blue is the operation performed in the previous step

Problem 150)

$Eq.1 \quad 8x - 4y + z = 1$

$Eq.2 \quad -2x + 10y - 3z = 5$

$Eq.3 \quad x + y + z = 5$

$Eq.1 \quad 8x - 4y + z = 1$

$Eq.2 \quad -2x + 10y - 3z = 5$

$Eq.3 \quad x + y + z = 5$

Count the number of variables

There are three different variables: x, y and z

$Number\ of\ variables = 3$

$Number\ of\ equations = 3$

$Number\ of\ equations \geq Number\ of\ variables$

The problem can be solved

$Eq.1 \quad 8x - 4y + z = 1$

$Eq.2 \quad -2x + 10y - 3z = 5$

$Eq.3 \quad x + y + z = 5$

Pick any of the three equations

We have arbitrarily decided to solve for z, but it is up to your free will to decide which one you want to start with

$Eq.1 \quad 8x - 4y + z = 1$

$8x - 4y + z = 1$

Solve for any of the three variables

Moves as positive

$8x - 4y + z = 1$

Moves as negative

It does not have to be in order, you can pick any of the remaining equations

$z = 1 - 8x + 4y$

$Solution\ 1 \rightarrow z = 1 - 8x + 4y$

By having an equation of z in terms of x and y we have reduced our problem to only 2 variables. Let's move to the next equation

Pick the next equation

$Eq.2 \quad -2x + 10y - 3z = 5$

$-2x + 10y - 3z = 5$

Replace the variable in red for the equation in Solution 1

$-2x + 10y - 3(1 - 8x + 4y) = 5$

Number in red is the operation performed in the current step

Number in blue is the operation performed in the previous step

$$-2x + 10y - 3(1 - 8x + 4y) = 5$$

Distribute the 3 inside of the parenthesis

$$-2x + 10y - 3 \cdot 1 - 3 \cdot -8x - 3 \cdot 4y = 5$$

$$-2x + 10y - 3 + 24x - 12y = 5$$

Combine elements with x

$$22x + 10y - 3 - 12y = 5$$

Combine elements with y

$$22x - 2y - 3 = 5$$

Solve for any of the two variables

$$22x - 2y - 3 = 5$$

Moves as positive

$$22x - 2y = 5 + 3$$

Moves as positive

$$-2y = 8 - 22x$$

By doing so, we are trying to reduce our system to only one variable

Moves as division

Apply:

$$\frac{A - B}{C} = \frac{A}{C} - \frac{B}{C}$$

$$y = \frac{8 - 22x}{-2}$$

$$y = \frac{8}{-2} - \frac{22x}{-2}$$

$$y = -4 + 11x$$

Solution 1 allows us to turn z into x and y
Solution 2 allows us to turn y into x
We have all we need to make an equation just in terms of x

$$Solution\ 2 \rightarrow y = 11x - 4$$

Pick the remaining equation

$$Eq.3 \quad x + y + z = 5$$

$$x + y + z = 5$$

First, let's replace that z for x and y

Replace the variable in red for the equation in Solution 1

$$x + y + (1 - 8x + 4y)$$

Number in red is the operation performed in the current step
Number in blue is the operation performed in the previous step

$$x + y + 1 - 8x + 4y = 5$$

Combine elements with x

$$-7x + y + 1 + 4y = 5$$

Combine elements with y

$$-7x + 5y + 1 = 5$$

$$-7x + 5y + 1 = 5$$

Replace the variable in red for the equation in Solution 2

And now let's replace y for an equation containing only x

$$-7x + 5(11x - 4) + 1 = 5$$

$$-7x + 5(11x - 4) + 1 = 5$$

Distribute the 5 inside of the parenthesis

$$-7x + 5 \cdot 11x + 5 \cdot -4 + 1 = 5$$

$$-7x + 55x - 20 + 1 = 5$$

Combine elements with x

$$48x - 20 + 1 = 5$$

Combine elements without x

Cha chin! An equation with only one variable

$$48x - 19 = 5$$

$$48x - 19 = 5$$

Moves as positive

$$48x = 5 + 19$$

Moves as division

We have our first variable! Now let's work backwards to get the other ones

$$x = \frac{24}{48}$$

Even after 150 problems we still have your back with fraction simplification!

$$\frac{24 \div 24}{48 \div 24} = \frac{1}{2}$$

$$Solution\ 3 \rightarrow x = \frac{1}{2}$$

Number in red is the operation performed in the current step
Number in blue is the operation performed in the previous step

209

Recall Solution 2

$$y = 11x - 4$$

Replace the variable in red for the value in Solution 3

$$y = 11\left(\frac{1}{2}\right) - 4$$

$$y = \frac{11}{2} - 4$$

$$Solution\ 4 \to y = \frac{3}{2}$$

Recall Solution 1

$$z = 1 - 8x + 4y$$

Replace the variable in red for the value in Solution 3

$$z = 1 - 8\left(\frac{1}{2}\right) + 4y$$

Replace the variable in red for the value in Solution 4

$$z = 1 - 4 + 4\left(\frac{3}{2}\right)$$

$$z = 1 - 4 + 6$$

$$Solution\ 5 \to z = 3$$

Multiplication of fractions:

$$\frac{A}{B} \cdot \frac{C}{D} = \frac{A \cdot C}{B \cdot D}$$

$$11\left(\frac{1}{2}\right) = \frac{11}{1} \cdot \frac{1}{2}$$

$$= \frac{11 \cdot 1}{1 \cdot 2} = \frac{11}{2}$$

Addition of fractions:

$$\frac{A}{B} - \frac{C}{D} = \frac{A \cdot D - B \cdot C}{B \cdot D}$$

$$\frac{11}{2} - \frac{4}{1} = \frac{11 \cdot 1 - 2 \cdot 4}{2 \cdot 1}$$

$$= \frac{3}{2}$$

Multiplication of fractions:

$$\frac{A}{B} \cdot \frac{C}{D} = \frac{A \cdot C}{B \cdot D}$$

$$8\left(\frac{1}{2}\right) = \frac{8}{1} \cdot \frac{1}{2} = \frac{8 \cdot 1}{1 \cdot 2}$$

$$= \frac{8}{2} = 4$$

Multiplication of fractions:

$$\frac{A}{B} \cdot \frac{C}{D} = \frac{A \cdot C}{B \cdot D}$$

$$4\left(\frac{3}{2}\right) = \frac{4}{1} \cdot \frac{3}{2} = \frac{4 \cdot 3}{1 \cdot 2}$$

$$= \frac{12}{2} = 6$$

$$x = \frac{1}{2} \ ; \ y = \frac{3}{2} \ and \ z = 3 \ »» \ \textbf{Answers}$$

<u>Systems of two or more variables: practice problems</u>

1) $3x + y = 1$; $2x - 2y = 4$ 2) $x - 4y = 5$; $2x - 6y = 0$

3) $x + y + x = 11$; $x - y + z = 2$; $x + y = 0$

4) $x + 2y - z = 2$; $x - y - z = -1$; $3x + y - 2z = -1$

Number in red is the operation performed in the current step
Number in blue is the operation performed in the previous step

Stay tuned on our YouTube channel and Instagram:

▶ User's Manual Academics

📷 @mathusermanual

for more help and solutions.

www.ingramcontent.com/pod-product-compliance
Lightning Source LLC
Chambersburg PA
CBHW082058210326

41521CB00032B/2479

* 9 7 8 0 5 7 8 8 9 6 8 2 3 *